建筑与园林的对话

缪朴建筑设计作品集

A Dialogue between Architecture and Landscape

Pu Miao's Architectural Design

给回忆中的母亲

建筑与园林的对话

缪朴建筑设计作品集

缪朴　著

A Dialogue between Architecture and Landscape

Pu Miao's Architectural Design

Pu Miao

同济大学出版社
TONGJI UNIVERSITY PRESS

目录
Table of Contents

III. 建筑局部采取园林的形态
Architecture Partially Taking the Form of Landscape

IV. 园林作为建筑中的城市公共空间
Gardens as Urban Public Spaces in a Building

前言
Introduction

本书的主题

笔者从 1990 年代初开始做探索性的建筑设计，20 多年来，有一个主题一直主导着探索的方向：如何将建筑与园林（或室内外空间）配对来服务于建筑功能。在本书中笔者选择了自己的 15 个设计作品，来演示到目前为止的心得。

为什么建筑与园林的关系对笔者有那么大的吸引力？这是因为建筑与园林的结合应是现代建筑在中国本土化的主要方向之一。

自从 1978 年开始改革开放以来，中国建筑师面临的挑战就是如何对工业化社会中发展成熟的"现代建筑"体系进行本土化。现代建筑包含两组成分：其中基本的原则（如理性的设计方法、形式由功能与当代技术决定、建筑要为社会大多数人服务等）应当为所有工业化中的社会所采纳；其中偶然的元素（如许多具体手法或风格）则只是现代建筑的诞生地（1920 年代以来的欧美）所留下的"胎记"。我们完全应当根据本地当下的生活习惯、审美口味、技术与经济的合适性等将其本土化 [1]。

中国人最喜爱的一个空间使用习惯，就是将建筑与园林（或室内外空间）配成许多组合。每个组合中的室内外空间强调彼此的沟通，共同为一项主要功能活动服务。这种多元环境使人在短距离之中，可以同时享受到室内的保护与舒适以及室外的开敞与自然生态。这一生活习惯在中国传统建筑中有结晶式的反映，决定了中国建筑的独特面貌：建筑通常被分解成许多小块，与同样被分解的室外空间组成布满整个基地的庭院建筑。具体形式可以是规整的四合院系列，也可以是布置更灵活的园林建筑。

这与欧洲建筑传统形成鲜明的对比。除去重心在南欧的古典时期外，欧洲建筑较多采取空地中耸立一个集中建筑实体的模式，建筑与周边绿地之间的界面强调分隔。现代建筑虽然在实体外形上做了不小变化，但在室内外空间的组织上大致传承了该手法。中西空间使用习惯上的这一不同可能反映了思想方法上的差异：中国文化倾向于混合本质不同的事物兼为我用，而欧洲人则似乎更强调对事物分类处理。另一个重要原因也可能与西欧气候较寒冷有关。

无论原因何在，建筑与园林配对这一空间使用习惯仍被今天的中国人所喜爱。近年来一些利用传统庭院住宅改建成的商业区（如成都的宽窄巷子）同时提供室内外餐座，颇受欢迎。即使是大部分住在高／多层公寓楼中的上海居民，也普遍要求客厅及厨房外有阳台 [2]。建筑与园林的配对同时为批判及修正我国建筑传统中的不足之处提供了便利工具。比如说，中国城市历来缺乏为市民日常生活服务的点式公共空间（如广场等），在城市化日益加快的今天甚至出现市民为抢占地方跳社交舞而发生纠纷的现象。如能在新建的公共或商业建筑中普遍插入小型公共庭园，将有助于缓解这一问题。

总之，选择建筑与园林结合作为现代建筑本土化的切入点，目的不在于复活表面上的传统形式，而是找出隐藏在表面形式下的基本空间结构及其服务的实际功能，按今天的中国社会现实加以取舍或调整，最后用当前的技术手段来创造出新形式 [3]。例如，全部建在地面上的传统庭院不再适合现代城市的高密度用地；同时，现代技术允许园林的位置从地面扩展到二层以上，绿化的布置从地面扩展到立面或头顶面。这些均迫使我们创造不同以往的室内外空间配对。又如，合院住宅中大家庭的几代人围绕一个庭院居住，已不再符合今天以核心家庭为主的社会现实，需要新的手法来探索在住宅面积小、城市用地紧的约束下兼有室内外空间的多户住宅设计。再如，传统庭院建筑与城市公共生活完全隔绝，这一倾向是否值得在今天的中国继续延续？

值得质疑。这就是为什么本书的主题被表述为"建筑与园林（或室内外空间）配对来服务于建筑功能"，而不是聚焦在"新四合院"之类的具体传统形式上。

建筑与园林的四种结合模式

本书的 15 个作品中，12 个已建成并均在专业刊物上发表过。每个设计除了建筑图纸及照片外，均有在初次发表或向业主汇报时写的文章一篇，用以说明主要设计概念（本书做了少量文字调整）。读者会发现，建筑与园林的配对始终贯穿每个设计，并与其他设计概念形成相辅相成的关系。

建筑与园林的结合有什么具体规律或手法可以遵循呢？笔者发现，国内外现对建筑—园林关系所做的研究本来就很少，其中个别企图总结设计规律的著作（如 Berrizbeitia & Pollak），又大致着眼在形式构图上，较少发掘建筑与园林的各种结合方式会如何独特地改善使用功能 [4]。恰恰是从这个更重要的角度出发，笔者在设计实验中总结出四种建筑与园林的结合模式。下面以本书中的作品为例加以说明。

1. 室外空间作为室内活动的替换场所

这是最广泛使用的建筑与园林的配对关系。如同在传统建筑总平面中通常会看到的，室外空间相邻与其配对的室内布置，可以与房间并列，也可围绕房屋或被室内空间所环抱，关键是人必须能很方便地在室内外之间转移，让室内的活动能溢出到庭院中，在一个更开放的有阳光、微风及花木的场所中展开。同时，室外空间的尺度、形状、铺砌、绿化、封闭程度等设计要符合所接纳的室内活动的功能要求。比如在上海闵行生态园接待中心中，会议大厅内的与会者可以到宽敞规整、视野开阔的临水平台上交谈，而客房里的休息者则能移座到绿化环绕的私密小庭院中。

2. 建筑与园林各自扮演功能配对中的独特角色

在这个结合方式中，建筑与园林利用其各自独特的环境特点，在一项功能活动中担当不同的一方职能。比如在宗教建筑（株洲朱亭堂、长兴寿圣寺扩建）中，身处房屋中的人很容易将"天然""纯净""生生不息"的自然景物看成神圣的境界，建筑则成为崇拜者居留的世俗领域。又如在展览建筑（上海新江湾城生态展示馆）中，园林可以成为展品，建筑成为容纳参观者活动路线的空间，参观者通过这个容器上的各种窗口观看作为展品的室外生态环境。再如在园林建筑（洛阳小浪底公园茶室及昆山思常路茶室）中，建筑被设计成独特的取景器，在不知不觉中诱使人进入特定的观景位置或路线，如从半沉在水中的位置来欣赏水景，或是从上到下逐步扫描一幅几层楼高的景观"立轴"，从而创造出对园林景观的独特体验。最后，思常公园餐厅中的条形庭院内种植攀援植物，使园林成为分隔餐座的绿色隔断。

采用各自扮演特定角色的方式来结合建筑与园林，两者之间的空间关系必须按功能活动的本质，从整个基地上做二维或三维的探索。这种对接将更不拘常规，在有些情况下两者之间可以只有视线相通。

3. 建筑局部采取园林的形态

在某些项目中，需要将建筑的局部（或整体但仅在大轮廓上）设计成接近园林景观的形式，从而使环境更好地烘托功能活动。比如说，建筑可能要减弱自己的人工化形式，以避免对周边的自然环境造成过大的冲击。在昆山风尚公园多功能建筑群中，建筑的一层被做成近似山坡的形态，与周围原来不够高大的现有土坡连为一体，改善了建筑体量与自然景观之间的平衡。要强调的是，这种变形不是庸俗地把建筑做成惟妙惟肖的"人造山水"，而是通过调整建筑的大致几何形状及材质，来提示它与自然景观基本形态的隐约联系。建筑作为工业产品的本质仍应得到忠实表达。这种变形只发生在局部，是因为我们仍然希望在基地上保留建筑与园林这两种不同质的环境。

4. 园林作为建筑中的城市公共空间

在新建的公共或商业建筑中加入与城市街道有直接联系的小型公共庭园，将有助于改善我国高密度城市中缺乏公共空间（特别是更适于社交的点式空间）的问题。昆山星溪公园游客中心中面临水池的屋顶广场被公众自发选中作为婚礼场所，说明城市居民确有此需要。不仅如此，此类公共庭园为周边建筑中的活动提供了比人行道更舒适的外延场所。如同昆山金谷园多功能建筑群所展示的，使用社区服务设施的居民可以在庭园中等候或集会，商店顾客可以在庭园中休息。因为这些园林空间与中国城市建筑密集的环境形成强烈的对比，它们又可能为社区建立新的地标。昆山里库袖珍公园及社区中心探索了这一可能。总之，在公共或商业建筑中设置许多小型公共庭园，将帮助这些建筑更好地与城市脉络沟通互动。

为方便阅读，本书按上述四种建筑与园林的结合模式在各个设计中是否起主导作用，将 15 个作品分成四组。但一个设计中往往会同时应用到多个模式，特别是"室外空间作为室内活动的替换场所"这一手法，因其直接丰富了人体的活动空间，本书中的大部分设计均使用了它。

建筑设计必须全方位地回答环境中的各种问题，所以除了建筑与园林（或室内外空间）结合的主题外，本书中的设计实验同时还对现代建筑本土化的其他课题做了探讨。比如在使用者进入建筑的空间序列上采用"逐步揭示"的手法，在建筑实体形式上借鉴传统建筑的"平缓的曲线"构图语言，创造各种用地效率高的"墙"来分隔不同功能区域，在构造上使用更适合发展中国家的建造技术与材料，在公共空间的设计上尝试切合我国高密度城市的"垂直功能分区""多层街道""硬地花园""硬质边界"等对策。在各个作品的设计概念文章中对这些其他探讨均有简要介绍。读者如需要对本书主题及上述其他设计概念的详细论证，可以在笔者的专业工作的另一半——学术论文——中找到 [5]。

回顾二十多年来的设计实践，笔者基本上是独自一人完成了各个工程的全部建筑设计工作（部分工程还包括城市、园林及室内设计），从方案、施工图到施工配合，乃至建筑摄影，无一例外。其中甘苦，只有自己知道。但这样做的好处是能将自己的理论研究心得真正贯彻到设计中，并落实到每个构造细节中去。自 1990 年代以来的这段时期，普通中小型建筑的设计如何在真诚体现中国的文化、经济与技术现实的前提下做创造性的探索，本书记录了笔者的个人尝试。

注释

[1] 缪朴，《用自己的声音说话——近作二则兼论"本土化"》，《建筑师》第 106 期（2003 年 12 月），第 20-28 页。

[2] 刘卫东、彭俊等，《上海市居民生活方式和住宅空间研究》（上海：同济大学出版社，2001），第 160-162 页。

[3] 笔者在确定传统建筑的基本结构特点（包括"室内外空间配对"）及其功能分析方面曾做过初步尝试，见：缪朴，《传统的本质——中国传统建筑的十三个特点》，《建筑师》第 36 期（1989 年 12 月）第 56-67 页、第 40 期（1991 年 3 月）第 61-80 页，以及《台湾大学建筑与乡研究学报》第 5 卷，第 1 期（1990 年 2 月），第 57-72 页。

[4] Anita Berrizbeitia and Linda Pollak, eds., *Inside/outside: Between Architecture and Landscape* (Gloucester, Mass.: Rockport Publishers, 1999). Jan Birksted, ed., *Relating Architecture to Landscape* (London: E&FN Spon; New York: Routledge, 1999).

[5] 可访问 www.pumiao.net。或用本书注释中提供的文献信息在 www.academia.edu 或 www.researchgate.net 中搜寻。

Abstract

How can architecture and landscape (or indoor and outdoor spaces) be integrated to serve building functions? This is the common theme underlying the 15 designs in this book, selected from this author's practice since the 1990s. Such an integration is believed to be one of the main directions to "localize" Modern Architecture in the context of China.

In the use of space, Chinese people love to combine buildings and gardens into pairs. In each pair, indoor and outdoor spaces are open into each other to form a mixed environment for an activity. Within a short distance, occupants can enjoy the indoor protection and comfort as well as the outdoor openness and nature. This habit in space usage has created the unique courtyard structure in traditional Chinese architecture, forming a salient contrast with the European convention of a consolidated building mass surrounded by greens. Such a habit is very much alive among modern Chinese.

Instead of reviving the traditional courtyards literally, however, the designs in this book experimented new forms that integrate architecture and landscape to serve the needs of today's Chinese society and urban context, with the use of contemporary technology. Four modes of integration between buildings and landscape can be concluded from the 15 designs.

1. Outdoor Spaces as Alternative Places for Indoor Activities
This is the most used integration. In this mode people must be able to move smoothly between the indoor and the outdoor spaces to allow indoor activities to be expanded into the garden setting.

2. Architecture and Landscape Each Playing a Unique Role in a Functional Pair
As demonstrated by examples in this book, building and garden can respectively accommodate the profane and the sacred spaces in religious architecture, the viewers and the exhibits in a museum, and the viewfinders and the views in garden buildings.

3. Architecture Partially Taking the Form of Landscape
For example, this integration can reduce the impact of the artificial building form, achieving a better balance between architecture and landscape in a project. The transformation aims at implying rather than faking the nature. It is partial so people still see two kinds of environment on the site.

4. Gardens as Urban Public Spaces in a Building
Chinese cities demand more nodal public spaces to serve residents' daily needs. Integrating small public gardens into new public or commercial buildings will not only relieve the shortage of public space but also allow building activities to spill into a setting better than the sidewalk. With its visual contrast to surrounding dense buildings, the public garden will also create a landmark much needed to improve the legibility of the crowded city.

More than one mode may be used in each design. In addition to the theme of integrating architecture and landscape, the 15 designs may also explore other concepts on localizing Modern Architecture to China. For detailed discussions of the thematic and other concepts, please see the other half of the author's work— the research papers (www.pumiao.net).

This author did all the architectural design work (plus urban, landscape and interior designs in some projects) of the buildings in this book, including schematic designs, construction documents, construction phase services, as well as photographing the completed buildings. How should we design small and mid-sized ordinary buildings creatively while embodying authentically the cultural, economic and technical realities of China since the 1990s? This volume records the author's personal experiments.

致谢
Acknowledgements

首先要感谢美国麻省理工学院建筑系的张永和教授为本书写的推荐语。本书得以出版，仰赖于《时代建筑》主编、同济大学建筑与城规学院支文军教授的帮助，并有幸得到同济大学出版社光明城的秦蕾、晁艳及张微所做的精心编辑。

笔者在本书记录的建筑实践中得到多位业主的支持及许多专业工作者（其中部分又是同学及朋友）的技术配合，他们的贡献均已在各个项目的"工程资料"中一一列出。特别要感谢的是上海市园林设计院的周在春及庄伟等，为笔者在国内的实践提供了最早的机会。同时要感谢昆山城市建设投资发展有限公司的周继春等多位领导及朋友对笔者的建筑探索长达十余年的支持。上海源规建筑结构设计事务所的张业巍为本书中的不少项目提供了精湛的结构设计，又是多年来陪笔者奔波于工地之间的好友。

最后要谢谢我弟弟缪林对书中照片所做的编辑。本书寄托了对母亲高丕琚的思念。

室外空间作为室内活动的替换场所
Outdoor Spaces as Alternative Places for Indoor Activities

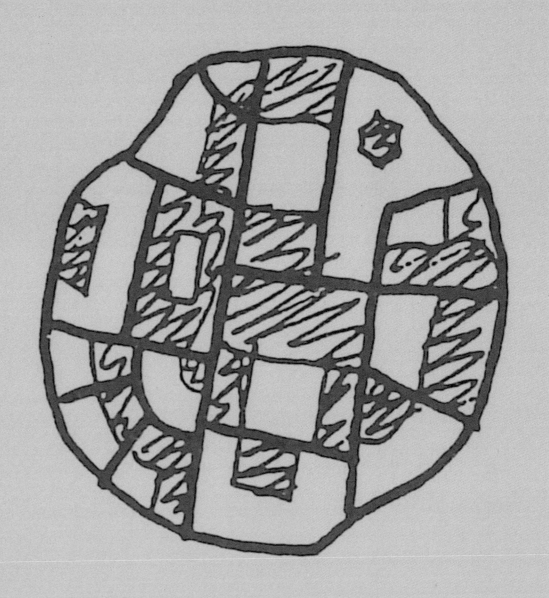

朴，《传统的本质——中国传统建筑的十三个特点》
插图之一，"室内外合作"。

"Pairing Indoor Spaces with Outdoor Ones," one of
the illustrations in Pu Miao, "Essence of Tradition—
The 13 Characteristics of Chinese Traditional
Architecture."

小河穿过的一组庭院
——上海闵行生态园接待中心（2004）
The Courtyards Strung by a Canal
—Reception Center, Minhang Ecological Garden, Shanghai (2004)

闵行经济技术开发区位于上海近郊，在 2000 年代初兴建了一座占地达 650 亩的公园"生态园"，给这个原本只有工厂住宅的区域增添了大量绿地。由于当地的许多公司及社区团体一直需要一个比办公楼或酒店更自然的场所，来举行小型团体的会议及休闲团聚活动，开发区决定在园中修建一个接待中心，内含公园管理楼及三个为上述招待功能服务的设施（后因经费限制调整为两个）。我有幸得到主持公园规划的周在春先生的邀请，在 2002 年参与了该项目的设计。

生态园沿南北向形成一个长条形，其间被东西向的城市道路隔成三个对外独立开门的单元。位于中段的名为"水生园"，因为园中保留了许多原有的鱼塘及小河。接待中心的基地被规划为一个与公园外形相似的长条形，从水生园东南角开始，沿东侧园墙向北延伸。

根据基地的形状，管理楼及两个会议设施"苇庄"及"幽僻处"依次由南向北，组成一个约 122 米长、37 米宽的长条形建筑群。每个设施内含会议大厅、餐厅、客房及厨房等。中心的三部分均有各自独立的入口，顺次开在建筑群西侧外墙上，面对一条与建筑群平行的车行道。

庭院传统的继承与批判

在当前占主导地位的西方文化影响下，业主原要求我们将本工程做成三个独立的小别墅，周围环以草地（图 1 左）。我们认为接待中心位于高密度城市的公园之中，基地周围的公园景观平坦开敞，节假日公园游人密集，独立别墅的做法将使该设施的室外空间因过分暴露而无法使用。为此我们借鉴了我国的传统庭院空间，将建筑分成小体块，将室外空间做成内庭院（图 1 右）。院墙在本设施与公园之间、本设施中的三个部分之间造成分隔。这一庭院式布局保证了人们能享受一种为中国人喜爱的休闲方式——在私密的环境中悠然观赏自然。

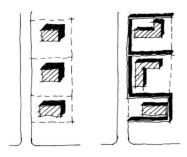

图 1. 独立建筑模式（左）与本方案（右）的比较。
Figure 1. Comparison between the freestanding-building solution (left) and our scheme (right).

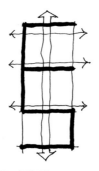

图 2. 视觉通道。
Figure 2. Visual channels.

传统庭院过分闭锁内向，反映了不鼓励人关注外部社会的传统文化。为纠正这一问题，我们在院墙上开启了各种洞口以形成视觉通道（图 2）。使用者能经常一瞥其他庭院以及外面的公园，从而扩大了空间感以及与城市公共生活的联系。主要视觉通道是贯穿长条形建筑群中央的一条南北向小河。它始于建筑群北端一眼涌泉，经过重重建筑及三部分之间围墙上的门洞，最后流入南端办公楼前面的一个大水池中。水道东西两面通过多道小桥联系。访问者走过任何一座桥时都可以通过沿小河形成的视觉通道一窥其他院落以至基地外的景色，但无法立即进入这些空间。沿基地的东西向设置了多条较短的视觉通道，始于建筑群西侧围墙上的狭窄豁口，通过庭院内部隔墙上的门洞或室内走道，最后终止在基地东侧一带保留下来的广玉兰树林中。与南北向的小河一样，这些视觉通道在不影响私密性的同时让使用者感受到外面公园的大片绿化。

墙的建筑

如果说在低密度环境中的建筑分区通常依赖房屋之间的距离（如大片草坪）来实现，本设计的庭院布局则大量采用了墙来实现这一目的。各种各样的墙体，像透明的花格墙或磨砂玻璃墙、上述带豁缝的院墙、芦苇"墙"，以及里面包着方形凉亭的圆弧墙，在狭窄的基地中不占用许多地方就能创造出多样的分隔。部分地方还采用了两道平行墙体内夹约 2.4 米宽室外空间的做法，来强调墙除了分隔以外的其他意义。如建筑序列开端的管理楼南立面正对城市道路，在这里，双重外墙表达了接待中心外墙既分隔又交流的主题。其中内侧外墙全部为玻璃，外侧外墙为开有各种洞口的砖墙，兼具景框及遮阳的功能。双重外墙之间的带状空间中设置了阳台、庭院等室外活动场所。在"苇庄"及"幽僻处"中，建筑体面临车行道的西侧面采用了另一种"夹心"结构，过渡空间被用作条形庭院（满足主要房间自然通风采光及景观需要）或服务房间。

室内外空间的配对

西方建筑传统中多将室内外空间分别处理成两个大块，与这种模式不同，中国人喜爱将每个主要房间都与一个室外空间配成对来服务于一个功能。本设计不仅采纳了这一模式，而且在现代建筑中发展了它。接待中心的建筑体大部被打散成多个小体块，与各种室外空间掺和在一起，形成多个配对关系，使各种生活功能都能在兼具人工与自然的场所中展开。即使是位于二层的房间也配有屋顶庭院，部分并能由此下到地面绿地。这些配对关系中的室外空间根据不同功能对环境的特定要求而采取了相应的设计，如在会议设施的入口处点缀竹林小院，在"苇庄"的会议大厅外设置宽敞规整的临水平台，而小河对过的客房则面对充满自然形态种植的小片绿地。

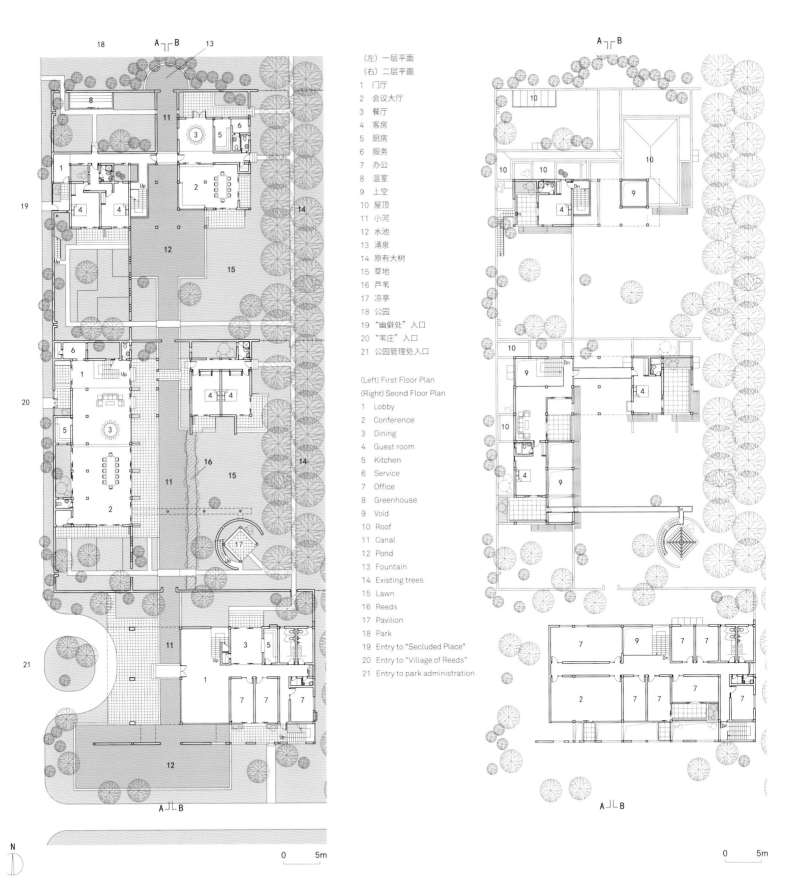

（左）一层平面
（右）二层平面
1　门厅
2　会议大厅
3　餐厅
4　客房
5　厨房
6　服务
7　办公
8　温室
9　上空
10　屋顶
11　小河
12　水池
13　涌泉
14　原有大树
15　草地
16　芦苇
17　凉亭
18　公园
19　"幽僻处"入口
20　"苇庄"入口
21　公园管理处入口

(Left) First Floor Plan
(Right) Seond Floor Plan
1　Lobby
2　Conference
3　Dining
4　Guest room
5　Kitchen
6　Service
7　Office
8　Greenhouse
9　Void
10　Roof
11　Canal
12　Pond
13　Fountain
14　Existing trees
15　Lawn
16　Reeds
17　Pavilion
18　Park
19　Entry to "Secluded Place"
20　Entry to "Village of Reeds"
21　Entry to park administration

庭院中的园林设计试图用几何形式及本地植物品种来表达上海郊区典型的水乡风景的基本结构，如通过将小河的各段做少许变形，使之在"幽僻处"中暗示鱼塘，而在"苇庄"中隐喻芦苇夹岸的水道。这道芦苇"墙"现在已长到预期的一人多高了，为"苇庄"中的会议厅与河东面更私密区域之间提供了所希望的视觉分隔。

普通的技术与材料

为了满足成本的限制，也为了探讨"现代建筑"的本土化，本设计着重采用了我国建筑工业目前在经济型低层项目中大量使用的材料与建造工艺，如在结构上使用了由砌体承重墙与部分钢筋混凝土框架（比单用框架时的截面要小）组成的"混合"体系。该体系的热工性能好，能满足本地区抗震设防烈度要求，允许一定的开洞。它还特别适用于本设计中有大量空间隔断的特点。其中部分混凝土框架被暴露出来，同时又成了立面构图中与大片白粉墙对比的构图元素。主要装修材料为白色涂料粉刷的内外墙面及天花、本色清漆木地板及木花架格栅，以及深灰色铝门窗，只在少数关键的位置画龙点睛式地采用了一些较精致的材料，比如像花架上的不锈钢攀藤钢丝及配件。我们甚至复活了一些被不少人认为是"过时"、但实际上价廉物美的传统做法，像"苇庄"及"幽僻处"两庭院中内隔墙上的砖花格。

总的来说，如果访问者通过当前流行的建筑模式来看接待中心，往往会找不到预期的东西。比方说，这里没有"主立面"或一二个最佳视角，因为建筑群的外观几乎就是一道白墙，对整个建筑的感受主要是通过人经过各个庭院时逐步积累起来的体验所形成的。这里也没有呼唤观众注意力的特异造型或新奇材料，因为我们想探索除了震惊以外其他感染人的途径。离竣工时去拍照要有多半年了，盼望着再次访问它。如果小河边的芦苇都已经长得那么高了，紫藤大概也快爬上花架了吧？（2005）

Abstract

The reception center of 4,400 square meters is located in a newly completed urban park in Minhang, a satellite town of Shanghai. The facility consists of a park administration building and two courtyard buildings for small-group conferences and retreats, "Village of Reeds" and "Secluded Place." The three parts are entered and used independently.

In contrast to the Western model of a consolidated building surrounded by lawn, the design adopts the Chinese courtyard concept to allow users to enjoy the garden in privacy. Avoiding the complete isolation in a traditional courtyard, a canal penetrates the three parts to form a north-south visual channel. Additional east-west channels are created through slits on layers of courtyard walls. These channels expand the sense of space and give users a glimpse of the public life outside.

Different from the use of lawns to separate functional zones in a low-density environment, the courtyard concept of this design uses a variety of walls, including the brick lattice wall, the "wall" of reeds, and the double facade, to create assorted separations within a tight site.

Essential to the courtyard concept, each major room in the facility, including those on the second floor, is paired with an outdoor space to create a mixed environment much loved by the Chinese people. The design of the outdoor spaces borrows typical patterns from local rural landscape, such as the reeds-bordered waterways and the fish ponds.

To meet a limited budget, but also to explore ways to localize modern architecture, the design focuses on the use of material and technology widely adopted by local building industry in low-rise, low-budget projects. We selected a hybrid structural system in which masonry bearing walls and site-cast reinforced concrete framing (slender than one used alone) work together. The system has a good thermal performance. It provides basic protection against earthquakes while still allowing for some large openings in facades. The design also revives a few traditional techniques, such as the brick lattice wall, which are being looked down as "outdated" but can actually create unique effects with little cost.

工程资料

地点 上海闵行经济技术开发区东川路（昆阳路口）
时间 2002—2004
建筑面积 4 400 平方米
业主 上海闵行联合发展有限公司
设计
建筑：缪朴（上海市园林设计院顾问）
结构：许蔓（上海市园林设计院，下同）
给排水：陈惠君
电气：周乐燕
公园规划：周在春

发表 / 获奖

《A+》（第 211 卷，2007），《Detail》（中文版，2006 年第 1 期），《建筑》（第 122 卷，2008），《建筑学报》（2005 年第 5 期），《时代建筑》（2004 年第 5 期），《Architecture Asia》（2009 年第 3 期）。
《Domus+78 中国建筑师 / 设计师》（2006）。
第六届远东建筑设计奖，佳作奖，2007 年。
Cityscape Architectural Review 建筑奖，入围，2006 年。
第一届深圳城市 \ 建筑双年展，2005 年。

Project Data

Location Dongchuan Road (at Kunyang Road), Minhang Economic & Technological Development Zone, Shanghai
Project Period 2002-2004
Floor Area 4,400 square meters
Client Shanghai Minhang United Development Co., Ltd.
Designer
Architecture: Pu Miao (Design Architect), Shanghai Landscape Architecture Design Institute (Architect Record)
Structure: Xu Man (Shanghai Landscape Architecture Design Institute, same below)
Plumbing: Chen Huijun
Electrical Engineering: Zhou Leyan
Park Planning: Zhou Zaichun

Publication/Award

A+ (Vol. 211, 2007), *Detail* (Chinese Edition, 1/2006), *Dialogue* (Vol. 122, 2008), *Architectural Journal* (5/2005), *Time+Architecture* (5/2004), *Architecture Asia* (3/2009).
Domus+78 Chinese Architects and Designers (2006).
Design Merit Award, the 6th Far Eastern Architectural Awards, 2007.
Short-listed, Cityscape Architectural Review Awards, 2006.
Exhibited at the 2005 First Shenzhen Biennial of Urbanism\Architecture, Shenzhen, China, 2005.

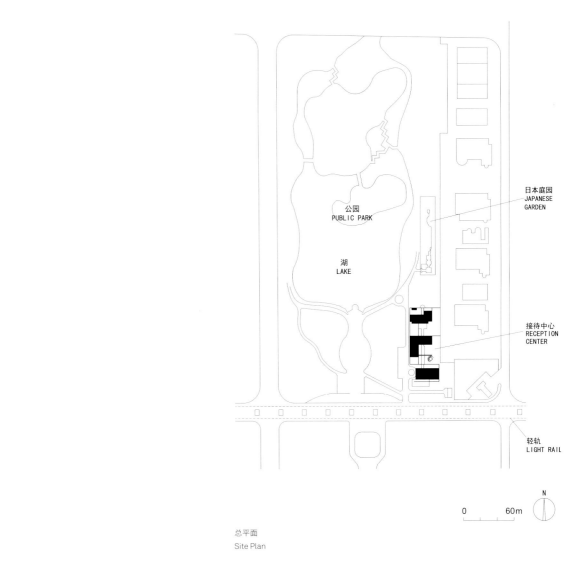

日本庭园
JAPANESE
GARDEN

公园
PUBLIC PARK

湖
LAKE

接待中心
RECEPTION
CENTER

轻轨
LIGHT RAIL

N

0   60m

总平面
Site Plan

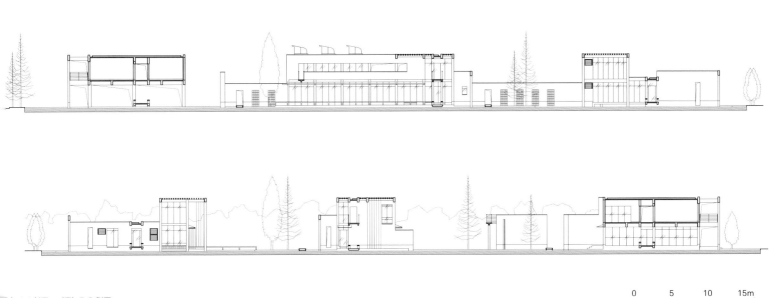

公园管理楼南立面，中部为作为南北视觉通道的小河。

South elevation of park administration; the canal as a visual channel is at the center.

公园管理楼南立面的双重外墙。

The double façade in the south elevation of park administration.

"苇庄"的主庭院，小图显示芦苇长成后景象。
Main courtyard of "Village of Reeds;" small picture shows the scene when reeds grow up.

"苇庄"的主庭院。
Main courtyard of "Village of Reeds."

鼓形墙内的凉亭与"苇庄"会议厅
上面的客房通过天桥相连。本图显
示芦苇尚未长成时。

A bridge connects the pavilion
behind the drum wall with guest
rooms above the conference room
in "Village of Reeds." This was a view
before the reeds grew up.

"苇庄"会议厅前的平台。
Platform in front of the conference
room in "Village of Reeds."

小河穿越"苇庄"的建筑，承重墙与混凝土框架的混合结构体系有利于降低造价。

Canal penetrates the building in "Village of Reeds." A hybrid system of bearing wall and concrete framing is used for the low-budget project.

（上左）"苇庄"会议厅室内，可见远处二层屋顶庭院。
(Above left) Interior of the conference room in "Village of Reeds;" an upper-level patio is visible.

（上右）从"苇庄"北部向南望。
(Above right) From the northern side of "Village of Reeds" looking south.

"苇庄"二层客房外的屋顶庭院。
Roof garden adjacent to a second-floor guest room in "Village of Reeds."

"幽僻处"入口。
Entry to "Secluded Place."

"幽僻处"入口庭院。
Entry courtyard of "Secluded Place."

"幽僻处" 会议厅内景。
Interior of the conference room in "Secluded Place."

"幽僻处" 的主庭院。
Main courtyard of "Secluded Plac

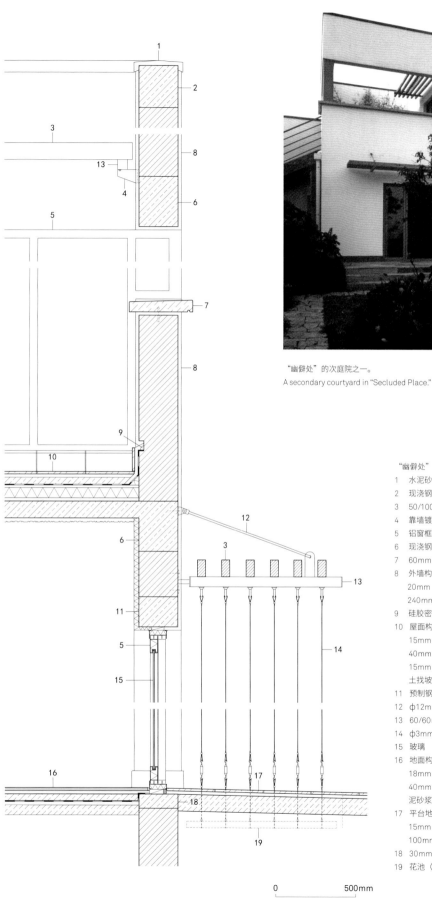

"幽僻处"的次庭院之一。
A secondary courtyard in "Secluded Place."

"幽僻处"客房翼南外墙垂直剖面
1　水泥砂浆压顶
2　现浇钢筋混凝土圈梁
3　50/100mm 防腐木花架条
4　靠墙镀锌钢支架支撑花架
5　铝窗框
6　现浇钢筋混凝土框架
7　60mm 花岗岩压顶，用金属销固定在墙上
8　外墙构造：
　　20mm 水泥砂浆粉刷内外两面，并做涂料；
　　240mm 砖墙与混凝土框架用钢筋拉接
9　硅胶密封屋面防水层尽端
10　屋面构造：
　　15mm 板岩 300/600mm；20mm 水泥砂浆；
　　40mm 配筋细石混凝土；油毡；屋面防水层；
　　15mm 水泥砂浆；60mm 硬质保温层；轻质混凝
　　土找坡；100mm 钢筋混凝土屋面板
11　预制钢筋混凝土过梁
12　φ12mm 镀锌钢吊杆及配件悬吊花架
13　60/60mm 镀锌钢管支架支撑木花架条
14　φ3mm 不锈钢缆及配件供植物攀援
15　玻璃
16　地面构造：
　　18mm 复合木地板及衬垫；20mm 水泥砂浆；
　　40mm 配筋细石混凝土；防水涂料层；20mm 水
　　泥砂浆；60mm 素混凝土；素土夯实
17　平台地面构造：
　　15mm 板岩 300/600mm；20mm 水泥砂浆；
　　100mm 素混凝土；素土夯实
18　30mm（至少）防潮层在所有墙体中设置
19　花池（远方）中埋金属管固定不锈钢缆

Vertical section through south wall of the gu
room wing in "Secluded Place"
1　Cement plaster coping
2　Site-cast reinforced concrete bond beam
3　50/100mm treated wood trellises
4　Galvanized steel wall bracket to support trell
5　Aluminum window frame
6　Site-cast reinforced concrete framing
7　60mm granite coping, anchored to wall by metal dov
8　Wall construction:
　　20mm cement plaster both sides, painted; 240
　　brick wall tied to concrete frame by steel rods
9　Silicon sealant to close the end of roofing felt
10　Roof construction:
　　15mm slate pavers 300/600mm; 20mm cem
　　mortar; 40mm reinforced concrete topping; buil
　　paper; roofing felt; 15mm cement mortar; 60
　　rigid insulation; lightweight concrete to fo
　　drainage slope; 100mm reinforced concrete slab
11　Pre-cast reinforced concrete lintel
12　φ12mm galvanized steel hanger and accesso
　　to anchor trellises
13　60/60mm galvanized steel tube bracket to sup
　　wood trellises
14　φ3mm stainless steel cable and accessories
　　plants to climb
15　Glazing
16　Floor construction:
　　18mm pre-finished wood floor panels with resil
　　pad; 20mm cement plaster; 40mm reinfor
　　concrete topping; liquid-applied waterpr
　　coating; 20mm cement mortar; 60mm un-reinfo
　　concrete; compacted soil
17　Terrace construction:
　　15mm slate pavers 300/600mm; 20mm cem
　　mortar; 100mm un-reinforced concrete; compac
　　soil
18　30mm (min.) moisture-proofing layer in all wa
19　Metal tube in planter (beyond) to anchor stain
　　steel cables

0　　　　　　500mm

"幽僻处"中的砖花墙。
e brick lattice wall in "Secluded Place."

砖花墙立面及水平剖面

1 240/115/57 斗砖二皮，外做涂料
2 240/115/57 眠砖一皮，两端伸入实墙，外做涂料
3 墙体构造：
   20mm 水泥砂浆粉刷内外两面，并做涂料；240mm 砖
   墙与混凝土框架用钢筋拉接

Elevation and horizontal section, brick lattice wall

1 240/115/57 brick, two rowlock courses, painted
2 240/115/57 brick, one stretcher course, ends extended
   into solid walls, painted
3 Wall construction:
   20mm cement plaster both sides, painted; 240mm brick
   wall tied to concrete frame by steel rods

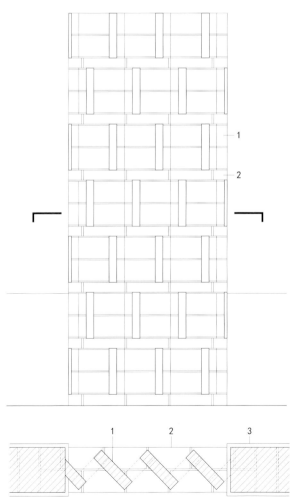

0   500mm

从"幽僻处"北部向南望。
From the northern side of "Secluded Place" looking south.

墙上的豁口形成东西向视觉通道。
ts on walls form an east-west visual channel.

小河北端由一个涌泉终结。
A fountain marks the northern end of the canal.

山地别墅的新类型
——苏州碧瀛谷别墅（2002）

A New Prototype of Hillside House
—House, Green Valley, Suzhou (2002)

本土化的切入点

对起源于欧洲的"现代建筑"进行本土化，其中一个最引人注目的课题就是室内外空间的关系。虽然西方建筑历史中不断有一些室内外结合的佳例出现，从古罗马的中庭住宅到弗兰克·盖里（Frank Gehry）的施纳贝尔（Schnabel）住宅（1989），但总的来说，西方人倾向于将一个建筑中的所有室内空间聚集成一个实体，周围环绕以基地中的室外空间。即使是在充满新概念的现代建筑史中，绝大多数创造发明也是集中在如何塑造或装饰这个"雕塑"的外形。与之相反，我国传统建筑中的室内空间总是被分割成多个单体，与各种室外空间混合在一起布满整个基地。生活中的每种主要功能通常都是在一个房间及一个与之配对的室外空间中同时展开的。这种配对关系可以采取多种几何形式，它可以很自由（如在园林中），也可以非常严谨（如在四合院中）。

中西不同的两种室内外空间组合模式都是符合人类需要的。但经验证明，无论是过去还是现在的中国人，似乎都更喜欢兼具人工与自然的"炒杂拌儿"式环境。人们希望能在房间里阅读工作，但只要走几步就可以到紫藤花架下就餐闲谈。事实上，我甚至怀疑这可能是一种对全人类都有吸引力的生活场所。只是由于西方文化在近现代建筑史中的霸主地位，大多数建筑师对这种非主流概念视而不见而已。所以，如何在今天的各类建筑中创造合适的室内外空间配对，是我近年来建筑实践的关注点之一。在此要强调的是，本土化不是简单地用现代技术来再现某个历史建筑形式（如四合院）。由于现代中国人的心理及文化背景已包含了许多新的因素，再加上建造技术在工业化社会中的发展，这都要求建筑师对配对的形式做创造性发展。以下用我在山地住宅中的尝试来说明这一过程。

第三种类型

在当代建筑实践中，造在山坡上的建筑通常有两种空间形式可选择：台阶型或塔楼型（图1左、中）。在台阶型中，建筑的各层沿坡地做退台式处理。典型的例子有工作室5（Atelier 5）在瑞士布吕格（Brügg）的瑞因帕克（Rainpark）住宅群（1970）。这样的组合不仅使每一层在其左右两个尽端都能直接接触地面绿地，如果基地的坡度足够陡的话，还可以将每层的屋顶做成露台或屋顶花园，与其上一层室内空间的地面取平，为房间中的居民提供近在咫尺的室外空间。这些都使台阶型不仅在外形上与基地结合得更贴切，而且也产生了更好的室内外空间配对。但这种与坡地的亲密关系同时也造成了它的缺点：台阶型通常无法产生醒目的建筑立面及高于一层的室内空间，部分背靠山坡的房间也难以实现对穿通风。

与之相反，塔楼型基本上是将平地上的一栋楼房搬到坡地上，唯一不同的是人们通过桥梁从顶层进入建筑。另外，其地面层可能局部悬挑或掉层。代表作有美国

图1. 山地建筑的三种类型：台阶型（左）、塔楼型（中）及复合型（右）。网点部分为与室内取平的室外空间。
Figure 1. The three types of hillside buildings: the terraces (left), the tower (middle), and the combined (right). Sand patterns are outdoor spaces that are at the same level of the indoor ones.

建筑师理查德·迈耶（Richard Meier）的道格拉斯（Douglas）住宅（1973）及瑞士建筑师马瑞欧·博塔（Mario Botta）的圣维塔莱河村（Riva San Vitale）住宅（1973）。金鸡独立的塔楼不仅外部造型鲜明，也便于建筑师在室内创造多种高度的空间。除一层外，建筑室内空间基本上没有其他机会与室外空间发生关系。（我国传统地建筑虽然外形像台阶型，但由于大多不用平屋顶，所以并没有实现现代台阶型全部优点。）

针对以上两种类型的优缺点，我探讨了第三种设计模式——复合型（图1右）。其基本原则是将建筑分成至少两个单体，其中一个做成台阶型，沿山坡从上到下展到另一个取塔楼型，树立在基地低端。塔楼形体中必须包含某种洞口，使它后面的筑望山下的视线得以穿过。通过利用台阶型中未用到的基地低端的上部空间，复合型弥补了台阶型外部造型及室内空间均平淡的缺点，同时又保证了绝大多数房间有远望景观并与室外绿地相邻。更重要的是，新类型的多个建筑体块可以被用来合出各种室外或半室外空间，从而在基地中创造出具有不同视野及私密性的室外环境，与各种室内功能形成更好的配对。图1中演示复合型的简图来自我在1991年美国加州伯克利某山地住宅做的设计，它说明了复合型的另一个有趣特点：该类中的室内外空间有可能形成两个系统，在三维空间中做成麻花状扭合。各块室外场虽处于不同高度，但总能在某处从建筑体中穿过，从而连通成一个空间体系。十年后，我和同事终于有幸在苏州碧瀛谷别墅区中实验了复合型的概念。

碧瀛谷案例介绍

占地约7公顷的碧瀛谷别墅区位于苏州西郊渔洋山的山麓，居高临下正对山面的太湖。我原为该住宅区设计了6种山地住宅类型，包括独家别墅、排屋及公寓等各种密度，业主只采用了其中两种。以下介绍的是基本上按原设计施工的一个类型。

基地呈正方形（约24.9米×24.9米），坡度在18°左右。入口车道在基地上方（侧）。建筑由两个均为三层的形体组成。位于基地东边的台阶型形体内含门厅、餐厅

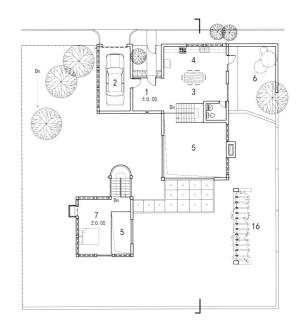

（下、中、上）　　　　　(Lower, middle, and upper)
一、二、三层平面　　　First, Second, and Third
1　门厅　　　　　　　Floor Plans
2　车库　　　　　　　1　Foyer
3　餐厅　　　　　　　2　Garage
4　厨房　　　　　　　3　Dining
5　上空　　　　　　　4　Kitchen
6　庭院　　　　　　　5　Void
7　客卧　　　　　　　6　Courtyard
8　客厅　　　　　　　7　Guest bedroom
9　屋顶花园　　　　　8　Living
10　通风井　　　　　　9　Roof garden
11　卧室　　　　　　　10　Ventilation shaft
12　平台　　　　　　　11　Bedroom
13　天桥　　　　　　　12　Terrace
14　书房　　　　　　　13　Bridge
15　水池　　　　　　　14　Study
16　花架　　　　　　　15　Pool
　　　　　　　　　　16　Trellises

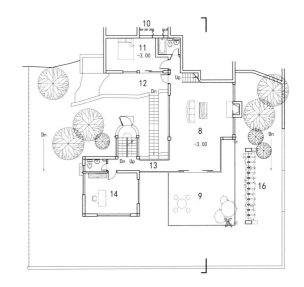

面　　　　　　Section
餐厅　　　　1　Dining
厨房　　　　2　Kitchen
客厅　　　　3　Living
屋顶花园　　4　Roof garden
卧室　　　　5　Bedroom
庭院　　　　6　Courtyard
花架　　　　7　Trellises

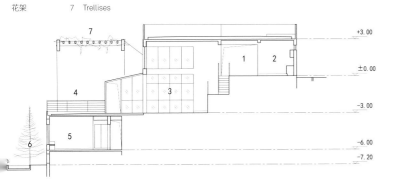

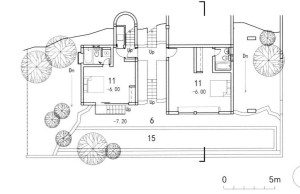

0　　　　5m

0　　　　5m

N

客厅及部分卧室，它从位于 -0.15 米的入口道路开始逐步跌落到 -7.20 米的底层庭院，在室外造型上产生两个台阶。其中上面一个成为与客厅取平的屋顶花园，下面的一个台阶则容纳了卧室外的底层庭院（庭院中的铺砌、水池及院墙均未按原设计建造）。为了改善台阶形体中靠山房间的通风，在北面卧室处设计了通风井。呈塔楼状的第二个形体耸立在基地西南角，包括了书房及另一些卧室。

两个形体之间留有一条豁口，一座天桥跨越其上。无论是在平面上还是立面上，两个形体均形成某种围合的关系。在平面上，这一围合界定出一块中心室外空间，类似于一个在山坡上展开的“四合院”。在建筑的主立面上，塔楼形体（包括它的花架）与台阶形体形成两个互相咬合的 L 形，两个 L 之间的空间正好是别墅公共生活中心的屋顶花园。从山下看，建筑呈现出一个独特的造型。一列台阶从天桥下穿过两个形体之间的豁口，将高处的中心室外空间与底层庭院连成一个室外空间系统。

这一复合型的空间结构为创造多样化的室内外空间配对提供了机会。首先，在别墅的入口处得以开辟一个小型的入口庭院（原设计的栅栏门未建），在开敞的室外道路与私密的室内门厅之间产生过渡，避免了西方建筑中常见的开门就进入“城堡”的突然跳跃。与门厅相邻的餐厅里，有可以打开通向建筑侧面室外平台的落地玻璃窗，在夏天可以让使用者在室外就餐。从餐厅步下客厅后，我们看到南面的整个玻璃墙可以打开，外面是别墅中最大的室外活动空间——6.5 米 ×5.0 米的屋顶花园。该空间正对太湖，顶上有遮阴花架，是延伸客厅内各种社交活动的最好场所。它同时还通过天桥为塔楼形体中的书房服务。除了以上公共性质的室外空间外，别墅中的大多数卧室也都有庭院与之配对。这些较私密的室外空间在建筑或围墙的遮挡下，都不会完全暴露在从山下往上望的视野中。如建筑底层的各个卧室开向带水池的院子。即使是位于台阶形体后部的卧室，也可享用上面说到的中心室外空间，虽然该空间中一大部分是台阶，但我可以想象它们会成为多种家庭活动的舞台。

本基地的最大优势是面对太湖，它给我们带来的第一个挑战是所有主要房间都必须能看到湖景。但第二个问题是这部分太湖景观本身实际上并没有多大变化。这就要求我们通过设计，用单一的原材料做出“一鸭三吃”的视觉宴席，使人们从一个房间走到下一个房间时能不断获得新的感受。而复合型的建筑布局对解决以上两个问题均有帮助。首先，客厅 4.8 米宽 ×5.4 米高的玻璃幕墙及其外面的屋顶花园使人们得以一览无余地眺望远方，形成别墅中湖面景观的高潮。与之相反，卧室、书房等则开以小窗，将湖面景观剪裁成各种尺度较亲密的画面（书房窗未按原设计的水平长条形施工）。底层卧室还辅以可坐的窗台，以强调在此地观景的平静心态。由于底层卧室高于外面的庭院，庭院围墙在保证室外活动私密性的同时不会影响从室内望湖。复合型中两个建筑形体之间的豁口，使位于基地北面的门厅、卧室及中心室外空间也能一瞥湖景。最后，由于台阶形体在室内还含有一个跌落，它使位于室内北端的餐厅也可穿过 6 米高的客厅上部看到太湖。

这个设计在本土化上的主要着眼点是建筑空间，但我们同时也试图在建筑硬体形式上探讨一种有别于西方建筑流行风格的构图方式。我在过去发表的一些研究中曾提出，无论是古代还是现代西方建筑立面，大多倾向于只含有曲线与直线正交两类形式，而且两者之间通常呈彼此对立甚至冲撞的关系；而在我国传统建筑、家具、舞蹈及书法形式中，我们往往会发现第三类构图元素，或我所谓的“平缓的曲线”[1]。汉字与罗马文字形式上的不同可以作为一个例子。这种从部分直线中延伸出来的曲线，有助于在以正交体系为主的总体构图中产生一种有变化但不失安静含蓄的效果。在本工程的南立面中，我们将屋顶花园东侧遮阴花架的支撑做成“平缓的曲线”，在两个 L 形体咬合的主题外增添了一个轻轻的次旋律。（2007）

Abstract

Hillside houses tend to belong to one of the two types, the terraces and the towe
While the tower has a more prominent visual appearance, the stepped profile
the terraces allows residents in the rooms same-level accesses to exterior space
This design explores a third type—the combined—that creates both a distinctiv
form and a better cooperation between indoor and outdoor spaces. The latte
distinguishes Chinese architectural tradition from the Western model of a cast
surrounded by greens.

Situated on the southern side of the mountains facing Lake Taihu, this 250-square
meter house in the Green Valley development, Suzhou, commands a panoram
view of the lake.

The house consists of two volumes. Containing the foyer, dining, and living room
the east volume assumes the terraced shape that produces a generous roof garde
in front of the living room. In contrast, the west volume which houses the stud
and guest room takes the tower form. The trellis roof of the tower extends over th
roof garden of the terraces, thus interlocking the two L-shaped volumes togethe
formally as well as functionally. A glass-enclosed bridge crosses the gap betwee
the two volumes.

A central outdoor space (or upper courtyard) is formed in the middle of th
embracing forms. Through the gap between the two volumes, a series of step
runs under the bridge and connects the upper courtyard to a lower one, forming
continuous sequence of outdoor spaces. The lower courtyard provides a garden fo
the bedrooms at the bottoms of both the tower and the terraces. The visual chann
along the gap allows rooms even at the northern side of the upper courtyard
have a view of the lake.

Instead of clashing irregular forms, the building composition experiments with th
use of a few “flat curves” that seem to grow out of the orthogonal form. Inducin
variety while remaining calm, this formal feature was first described in one of m
previous studies on Chinese traditional architecture.

工程资料
地点 江苏省苏州市吴中区环太湖大道碧瀛谷住宅区
时间 2001—2002
建筑面积 250 平方米
业主 上海华鑫建设发展有限公司
设计
缪朴设计工作室（缪朴、罗继润）；江苏省纺织工业设计研究院。室内设计由其他设计师提供。

发表
《建筑学报》（2007 年第 11 期）。

Project Data
Location Green Valley development, Huan Taihu Avenue, Wuzhong District, Suzhou, Jiangsu Province, Chin
Project Period 2001-2002
Floor Area 250 square meters
Client Shanghai Huaxin Construction and Development Co., Ltd.
Designer
Miao Design Studio (Design Architect), Pu Miao, Luo Jirun; Textile Industry Design and Research Institute
Jiangsu Province (Architect of Record). Interior design was by other parties.

Publication
Architectural Journal (11/2007).

注释
[1] 缪朴，《用自己的声音说话——近作二则兼论“本土化”》，《建筑师》第 106 期（2003 年 12 月），第 27 页。

建筑南立面（底层庭院围墙及水池未建）。
South elevation (the wall around the lower courtyard
and the pool were not built).

碧濂谷住宅区总平面		Green Valley Master Plan
1	本住宅基地	1 Site of this house
2	住宅区入口	2 Estate entrance
3	旅馆区	3 Hotel
4	旅馆服务区	4 Hotel services
5	网球场、车库	5 Tennis court, parking
6	台阶式公寓	6 Terraced apartments
7	会所	7 Club
8	登山步道区	8 Hiking
9	环太湖大道	9 Huan Taihu Avenue
10	太湖湖滨	10 Waterfront of Lake Ta

0   50m   N

从客厅望餐厅。
From living room looking upward into dining room.

餐厅及相邻室外边院。
Dining room and the adjacent side yard.

门厅的条窗可一瞥远处湖景。
t in the foyer allows a glimpse
ard the lake.

客厅开向屋顶花园及太湖。
The living room opens into the roof garden and Lake Taihu.

客厅外屋顶花园。
The roof garden in front of the living room.

从二层北部卧室穿过两个建筑形体之间的天桥望湖。卧室外为中心室外空间。

From the bedroom at the northern side of the second floor looking through the gap between the two building volumes toward the lake. The bedroom opens into the central outdoor space (upper courtyard).

在天桥下望中心室外空间。

From below the bridge looking toward the central outdoor space (upper courtyard).

从中心室外空间穿过天桥望湖及底层庭院。

From the central outdoor space (upper courtyard) looking beyond the bridge toward the lake and the lower courtyard.

从建筑南面的底层庭院望两个建筑体之间的豁口。

From the lower courtyard on the south looking through the gap between the two building volumes.

层的一个卧室。

edroom on the first floor.

建筑西面望被建筑单体围合的中
庭外空间。

central outdoor space (upper
rtyard) embraced by the two
ding volumes, as seen from the
t side of the building.

"墙"中世界
——昆山阳澄湖公园游客中心（2012）

A World in a "Wall"
—Visitor Center, Lake Yangcheng Park, Kunshan (2012)

江苏省昆山市历史悠久，改革开放后又变成一个飞速成长的工业城市。城市建成区逐步向东西方向延伸，西端最终扩展到阳澄湖畔。为了避免湖滨全被小别墅开发所瓜分，保证中心城区的普通市民能享受市郊最大的水岸，本工程的建设单位昆山城市建设投资发展有限公司于数年前开始在湖滨开发昆山阳澄湖公园。该开放式公园占地 42.5 公顷，位于城市东西向主干道马鞍山路在湖滨的端点，是昆山在湖滨最大的一片公共绿地。目前公园已基本建成，每逢周末游人众多，有在湖滨广场放风筝的，还有来骑自行车的。

公园的主体部分基本呈一个朝湖面开口的月牙形平面。园内主要步行道为一条与月牙基本平行的弧形木栈道，从西南方的湖滨广场开始，穿过全园中部，再回到西北方的水边，沿路将园内的主要景观串联起来。在木栈道离开湖滨广场不远处，是公园最大的一片游客停车场。停车场北边与木栈道之间的一块条形绿地，就是本工程的基地。

本工程的建筑面积不到850平方米,它的设计应是一篇节制但不缺理念的小品文。理念必须从最根本的地方去找。游客中心的本质是一个地方的大门，但要有门就必须先有墙。考虑到基地正好是线形的，为什么不把本建筑就做成一道"墙"呢？人们现在从车上下来，会看见沿停车场整个北边沿是一道 99 米长、7.75 米高的"墙"。"墙"上开了三个大小不一的门洞，提供了进入公园的通道。在周边墙面的衬托下，这些门洞中揭示的风景显得特别诱人。

但人们马上会发现这不是一道简单的实墙。在南面，它由里面的混凝土墙及外面与之平行的一层木格栅组成（木格栅与混凝土墙之间有若隐若现的钢楼梯）。木格栅及其上将长成的攀援绿化会使这道"墙"显得轻盈多变。格栅墙在建筑西端突然偏离后面的混凝土墙，形成一个独特的终端造型，向附近的湖滨（从停车场看不见）致意。就连外墙尽端的细部设计也体现了这一多层表皮的概念。

最大的惊奇发生在人们进入门洞时。向东西方向一望，他们会发现在这道"墙"的南北两面混凝土外壳之间藏着复杂多变的景观。由于建筑内部没有多少横隔墙，游客可以看到沿着建筑内部的长向罗列着各种空间，包括明亮与幽暗的、室内与室外的、高与低、水平与斜面的，等等。比如说，坡道下的下沉庭院让人联想起一个树木林立的浓荫峡谷。这些都是通过在狭长空间中交错布置房间与庭院、水平楼板与坡道，外加天窗与其他垂直流动空间来形成的。

一层包含一个为游客导游的展览空间及多个商店。人们通过坡道可以上到一个坐落在茶室与一个商店之间的公共屋顶庭院。茶室西面有另一个自用的屋顶庭院，是望湖的最好地方。无论是在地面还是二层，每一个室内空间都与一个或多个室外空间配成对来服务于一个建筑功能，这是中国传统建筑的基本特点之一，而现代中国人也仍然喜爱这种室内舒适与户外自然均近在咫尺的复合环境[1]。

人们通过三个门洞穿过"墙"后，将发现一个幽静的木平台把建筑背面与现的木栈道连接起来。基地上原有的树木大部分被保留下来，穿插在平台中。刚来将离去的游客可以从平台边的商店要一杯啤酒，在浓密树荫笼罩下的平台上享用。

可以说，这个不大的建筑设计完全是围绕着这样一个最终目的：让游客像剥洋那样不断在一道"墙"中发现新的东西。这种通过让人走过不同空间来逐步揭示筑面貌的做法，与不少西方建筑师（或他们在国内的模仿者）常用的以一个主要部造型来立即"震撼"观众的做法正好相反。陈从周先生曾引用杜甫"一丘藏曲折缓步百跻攀"来说明我国的造园原理之一[2]，这也正是本设计学习我国建筑传统的要着眼点。（2014）

注释

[1] 缪朴，《传统的本质——中国传统建筑的十三个特点》，《建筑师》第 36 期（1989 年 12 月）第 56-67 第 40 期（1991 年 3 月）第 61-80 页，以及《台湾大学建筑与城乡研究学报》第 5 卷，第 1 期（1990 年 2 月第 57-72 页。

[2] 陈从周，《说园》（上海：同济大学出版社，1984），第 6 页。

Abstract

Lake Yangcheng Park of 42.5 hectares is the largest public green space on the la shore of Kunshan. The linear building site lies between the main public parking and a curved boardwalk which leads people to the waterfront and other attractio

A visitor center is the front door of a place. But a door presumes the existence a wall. Considering the linear site, it is appropriate to make the building a "wa Indeed, people entering the parking lot see a curved "wall," 99 meters long and 7 meters high. Three doorways on the "wall" open up passages into the park. T sceneries revealed via the openings appear particularly enticing.

But this is not a simple solid wall. A layer of wood trellises parallel to an inr concrete wall makes the "wall" appear light and ever-changing. At the west end, t trellis layer deviates from the concrete wall and forms a unique gesture, nodding the nearby lake shore.

The big surprise comes when people enter the doorways. Looking east and west, th will find that few solid crossing partitions exist inside the "wall" between its two concr shells. Various spaces—bright and dim, indoor and outdoor, high and low, orthogonal a diagonal—are juxtaposed along the longitudinal direction of the cavity.

Most rooms in the building can spill their activities into one or more adjace outdoor spaces. The pairing of indoor and outdoor spaces is one of the fundamen characteristics of Chinese traditional architecture.

程资料

点 江苏省昆山市环湖路阳澄湖公园

间 2009—2012

筑面积 844 平方米

主 江苏省昆山城市建设投资发展有限公司

计

筑：缪朴设计工作室（缪朴）；汉嘉设计集团（蒋宁清）

构：上海源规建筑结构设计事务所（张业巍、刘潇）

备：汉嘉设计集团（穆立新、于洋、吴秋燕）

表

新建筑》（2014年第4期），《Architectural Review Asia Pacific》（2015
6-7月刊），《FuturARC》（2016年1-2月刊），archdaily.com。

oject Data

cation Lake Yangcheng Park, Huanhu Road, Kunshan, Jiangsu
ovince, China

oject Period 2009-2012

oor Area 844 square meters

ent Kunshan City Construction, Investment and Development Co., Ltd.

signer

chitecture: Miao Design Studio (Design Architect), Pu Miao; Hanjia
sign Group, Shanghai (Architect of Record), Jiang Ningqing
ucture: Shanghai Yuangui Structural Design Inc., Zhang Yewei, Liu Xiao
gineering: Hanjia Design Group, Mu Lixin, Yu Yang, Wu Qiuyan

blication

w Architecture (4/2014), Architectural Review Asia Pacific (Jun.-
../2015), FuturARC (Jan.-Feb./2016), archdaily.com.

费尔蒙酒店
FAIRMONT
HOTEL

阳 澄 湖 公 园
LAKE YANGCHENG PARK

马鞍山西路 WEST MAANSHAN ROAD

阳 澄 湖
LAKE YANGCHENG

阳 澄 湖 公 园
LAKE YANGCHENG PARK

苗木繁育基地
NUSERY

游客中心
VISITOR
CENTER

广场 PLAZA

环湖路

HUANHU ROAD

住宅施工区
HOUSING CONSTRUCTION SITE

0   100   200m

N

总平面
Site Plan

从停车场看建筑南立面西段，右面为中部门洞。
The west part of the south elevation, as seen from the parking; the middle doorway is to the right.

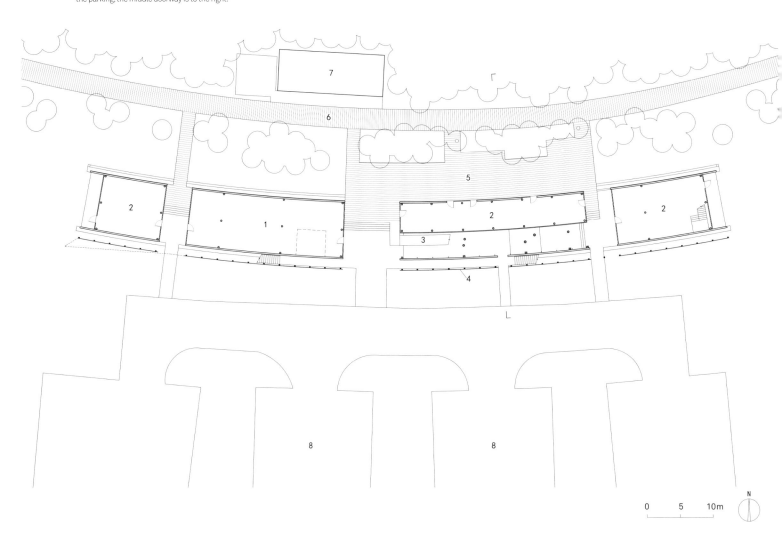

停车场看建筑南立面东段，左面为中部门洞。

east part of the south elevation, as seen from
parking; the middle doorway is to the left.

（左）一层平面		(Left) First Floor Plan		（上）二层平面，		(Upper) Second Floor Plan,
展览	1	Exhibition		（下）屋顶平面		(lower) Roof Plan
商店	2	Shop		茶室	1	Cafe
坡道	3	Ramp		厨房	2	Kitchen
格栅墙	4	Trellis wall		厕所	3	WC
平台	5	Deck		上空	4	Void
现有栈道	6	Existing boardwalk		屋顶庭院	5	Roof court
现有厕所	7	Existing WC		商店	6	Shop
公园停车	8	Park parking		格栅墙	7	Trellis wall
				屋顶	8	Roof
				花架	9	Trellises
				天窗	10	Skylight

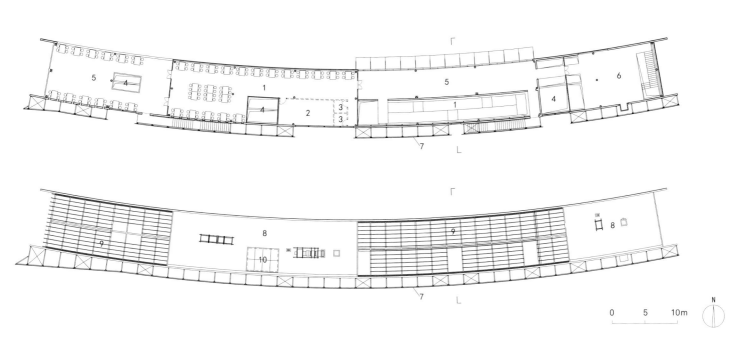

0 5 10m

N

通过西部门洞北望。
Looking north through the west doorway.

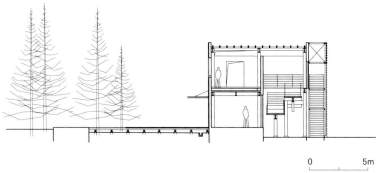

剖面
Section

0   5m

面中木格栅的西端。

t end of the trellis wall in
south elevation.

（上）建筑西立面向湖打开自己。
(Above) West elevation of the building opens itself toward the lake.

（下左）木格栅西端独特的终端构架。
(Below left) The unique framing behind the west end of the trellis wall.

（下右）木格栅与混凝土墙之间的空间。
(Below right) The space between the trellises and the concrete wall.

从中部门洞内东望。
Looking east from the middle doorway.

部门洞内沿无横隔墙的建筑内部西望，可见采
及远处湖景。

n the middle doorway looking west along the
ior of the building, no crossing partition hinders
views of a lighting well and the distant lake.

沿建筑北边及坡道展开的商[店]
沿无横隔墙的建筑内部长[向远]
望，可见远处室外绿化。

Shop along the north side a[nd]
the ramp. No crossing partit[ion]
hinders the longitudinal view [to]
the distant greenery beyond [the]
east end of the building.

（下左）坡道下的下沉庭院。
(Below left) The sunken cour[t]
under the ramp.

（下右）坡道。
(Below right) The ramp.

东部门洞内东望，可见多变
空间及建筑东端外的绿地。

ed interior space and the
enery beyond the east end
e building, as seen from
east doorway looking east.

由坡道进入的中部屋顶庭院。
The middle roof court entered from the ramp.

从茶室（尚未装修）东望中部屋顶庭院及坡道。
From the cafe (to be finished) looking east toward
the middle roof court and the ramp.

（上）从茶室（尚未装修）内北望室外，右为采光井。
(Top) Looking north from the interior of the cafe (to be finished), with a lighting well to the right.

（中）可瞭望湖景的西部屋顶庭院。
(Middle) West roof court overlooking the lake.

（下）从二层东端的商店西望。
(Bottom) From the shop at the east end of the second floor looking west.

建筑北面的木平台。
Wood deck along the north side of the building.

复合建筑
——昆山阳澄湖公园景观建筑（2010）
Multivalent Architecture
–Pavilions in Lake Yangcheng Park, Kunshan (2010)

建筑设计界目前流行一种做法，将一个复杂庞大的工程设计归结到一个与使用功能或基地无多少关系，但外观独特的符号形象上。这种假大空的做法规避了特定环境中人的复杂功能需要，通常在形式上也缺少真正的创意。西方有些建筑评论家恰当地称之为"一句话"建筑（"One-liner"）。在我国，某种程度上建筑成为权力与资本的广告牌，这种现象也并不少见。

"一句话"建筑的对立面是什么呢？英国建筑评论家查尔斯·詹克斯（Charles Jencks）在他的经典著作《现代建筑运动》中开宗明义地宣告了他认为的好建筑，其形式应同时能让人从多个层面上来解读，如功能、美观、建造技术、基地、成本等。但是在每个层面中，也能通过创造性的折中来同时容纳不同的侧重点。詹克斯针对这种建筑总结道："我们一次次重访它们，不一定是为了它们的某一种特定意义，更是因为多种意义以动人且深厚的方式交织融合成一个有力的组合。我将这种特性称之为多重意义……"[1]。

如柯布西耶的马赛公寓在有限的基地中容纳了三百多户家庭以及公共服务设施，在创造舒适的家园与限制城市蔓延之间取得了平衡。再以公寓阳台边缘的一个L形牛腿为例，詹克斯发现它可以同时被看作桌子、栏板结构支撑、杂物储柜、建筑外面上的一个构图元素等。绝大多数成熟的建筑体系（如我国的传统建筑）都具有丰富的复合意义。以下介绍一下我在一个景观建筑的设计中创造复合意义的心得。

江苏省昆山市在最近20年中从一个江南水乡古镇演变成一个新兴的工业城市，建成区的西端现在已经延伸到阳澄湖边。这里新建了一个公园，将成为城市在湖边的唯一一块公共绿地，以满足公众与附近酒店顾客休闲的需要。公园中东面一个方形区域中要求建造三个服务设施（B、C、D栋）。由其他设计公司负责的公园景观设计已经采用了多种对比强烈的几何形式。为了避免加上建筑后整个公园的环境形式过于喧闹，我们提出让这三个建筑共用一个形式母题及结构体系，同时在每个单体中再根据周边环境及朝向等做局部变化，局部调整只涉及隔断等非结构构件。由于三个建筑的功能完全一样，它们远看相似的外貌有助于游客尽快发现他们需要的服务，但当游客走近各个建筑后，会发现每个单体实际上各自不同。目前D栋已建成。

室内外空间的复合

西方园林中的建筑多被设计成被绿化环绕的实体，实体内包含的主要是室内空间，18世纪英国风景式园林（Landscape Gardens）中的园亭（Temple）是典型的例子。与此不同，我国的传统园林建筑大多为同时包含室内外空间的庭院式组合。我们在实践中发现，我国当代居民大多仍旧喜爱庭院，因为它能在一个小范围内让人同时享受到室内外空间各自的优点，像室内的舒适方便与室外的亲近自然。我们因此在本设计中尝试用现代手法来重现这样的室内外复合空间。

实现以上目标的策略是在布置平面时，不再先入为主地把室内外空间当成两个割裂的大类来处理，如仅仅在建筑的"外面"布置露台。粗看一下这三个36米×14米的平面，好像都是由传统的房间与走廊组成，但实际上这些"房间"中既有全封闭的室内空间，也有有顶无墙的半室外空间，更有一大部分是头上只是花架的室外空间。三个室内空间（出租、小卖部、公共厕所）也不再被捏合成一块，而是按最方便游人使用的方式与多个室外休息空间交错布置。为此我们选用了大跨度刚架结构体系，平面中没有任何柱子，使我们得以在三个平面中根据不同的人流及景观，自由布置形成室内外空间的非承重隔墙与屋顶。

我们设想大多数公众在这里需要的是一个可以随便坐坐、有绿荫而又不必付费的场所，所以将建筑中心最大的一个室外"房间"保留给这一功能。多项研究证明，提供饮食有助于搞活公共空间。我们因此在该休息空间靠近建筑外公共空间（如D栋旁的椭圆广场）的角落上，布置了一个出售冷饮茶水的小卖部。全室内的出租房间将用作咖啡厅或礼品店，相信其中的活动也会蔓延到周围的室外空间。

多功能外墙

这个兼容多种空间的长方形体又是通过复合外墙来界定的。建筑的南、北立面各由一道36米长、3.34米高的双重玻璃外墙构成。本建筑中采用的刚架是钢格构式的，其中梁、柱部分均由中距500毫米的双肢钢管组成。这一500毫米间距正好成为双重外墙的中空空间。现代版的双重玻璃外墙从20世纪90年代开始就在西方被广泛使用，但基本上局限在高层办公楼中，使用的目的也集中在节约能源上。在本工程的应用中，我们尝试给它增加两种新的意义。

首先，我们在双重外墙中设计了二层搁板，使这个中空空间可用来生长盆栽植物。这两道"绿色"外墙与室外绿化相互影映，将突显建筑身处公园中的特点。出租中的商店或咖啡厅也可利用部分中空空间陈列它们的商品。如是内装有色液体的玻璃器皿之类，在室外阳光投射下将特别吸引室内顾客的注意。

其次，我们在内外墙体上均使用了磨砂玻璃，但在特定位置不做磨砂，留出透明的局部。这些"窗口"在内外层上的位置又被错开。无论是从室内或室外看去，该外墙体系呈现出至少三个层次的通透感：半透明的外层、在透明外层后面的半透明内层、以及内外层均透明的洞口。当观众移动时，这些窗口还会彼此掩映。加上中空中的植物，创造出多变的视觉形象。由于磨砂图案在三栋建筑中各不相同，这一设计也为每个单体提供了可识别的个性。

与此同时，我们采取了好几个设计措施来完善双重玻璃外墙的节能功能。这些措施考虑了我国中小型建筑的施工和管理的现实条件，尽可能简单易行，不需依靠计算机控制的马达或光电感应之类的高技及高成本手段。比如说，两道外墙的外层上部与内层的全部都是可开启的。昆山的气候温和但四季分明，在室外气候宜人的春秋季可打开双层上的窗，充分利用自然通风。在冬季可同时关闭内外层上的窗，由于内外层均采用保温玻璃（外层还用了 LOW-E 镀层），使中空空间成为热传导的一道阻碍。在炎夏可只开启外层窗，利用上下热压差在中空空间中产生垂直通风，同时降低对室内气温的影响。为了保证空气的垂直流动，中空中的搁板只覆盖了 2/3 的空腔平面。B、C 栋中的空调室内空间均位于建筑北侧，其南立面外有大片花架遮阳。D 栋的出租房间虽位于南侧，但室外有许多高大树木。我们同时将该栋南立面的搁板设计为有助于遮阳的木板（暂未实施）。

本建筑建成后，我们在外墙中还发现一个没有预料到的视觉现象。从室内向外看，在透明玻璃后面或贴在磨砂玻璃上的花草叶片与远离磨砂玻璃的植物会产生如同水墨画中的干、湿笔对比，前者深黑明晰，后者如淡墨晕染。我们祈愿日后的建筑使用者能让这些新购置的绿萝逐渐长大，让它们能越过搁板架向上下蔓延，使整个外墙更像一幅绿色的画。绿色建筑发展到今天，其科学及道德上的意义已经得到全球无论贫富社会的共同认可。现在它应当超越只在技术或法规上寻求解决办法的局限，使自己与建筑其他方面形成相得益彰的有机组成部分，向更为广阔的方向寻求复合性的创新。（2011）

注释

[1] Charles Jencks, *Modern Movements in Architecture* (Harmondsworth, UK: Penguin Books, 1985), p.14.

Abstract

With the western suburb of Kunshan reaching Lake Yangcheng, the city constructed a park on the lakeshore, which is also the only large public space there. In the eastern rectangular area of Lake Yangcheng Park, three pavilions (B, C, D) were planned to serve users' resting, snacking and other recreational needs. To avoid competing with the already diverse landscape forms, the design proposes a simple prototypal form and structural system shared by the three pavilions, while their non-structural elements can be adjusted to local conditions. Pavilion D has been completed.

The design tries to make the seemingly simple building a multivalent creation, with each element simultaneously serving several purposes. Inside the 36-by-14-meter building area, indoor, outdoor and semi-outdoor spaces are mingled to afford users close touches with the comfort of a room and the refreshment of nature. This can be seen as a continuation of Chinese courtyard architecture, avoiding the Western tradition that often makes park buildings solid objects surrounded by greens. A long-span and modular structure is used to allow for free placements of partitions and roofs, better integrating indoor and outdoor spaces.

The south and north elevations of the pavilion are made of a double-skin glass façade. By opening/closing windows on either or both skins, the façade can passively ventilate or insulate the room according to different seasons. Apart from its traditional energy saving function, however, the design adds two new meanings to the double-skin facade. First, shelves are inserted in the cavity (covering only 2/3 of it) to grow potted plants or to display the merchandize of a shop. Secondly, sand-blasted and transparent areas are created on the glass. The transparent "windows," located differently on the two skins, generate varied overlapping effects, making the wall appear to have at least three depths. Thus the multivalent approach pushes green designs to go beyond technical and regulatory concerns and to be better integrated into the broader agenda of architecture.

工程资料

地点 江苏省昆山市马鞍山西路（环城西路与环湖路之间）
时间 2009—2010
建筑面积 160 平方米
业主 江苏省昆山城市建设投资发展有限公司
设计
建筑：缪朴设计工作室（缪朴）；汉嘉设计集团（蒋宁清）
结构：上海源规建筑结构设计事务所（张业巍、刘潇）
设备：汉嘉设计集团（郭忠、于洋、吴秋燕）

发表

《绿界》（《Domus 中文版》别册，2011 年 10 月），《时代建筑》（2012 年第 4 期）。
《当代中国建筑地图》（2014）。

Project Data

Location West Maanshan Road (between West Huancheng Road and Huanhu Road), Kunshan, Jiangsu Province, China
Project Period 2009-2010
Building Area 160 square meters
Client Kunshan City Construction, Investment and Development Co., Ltd.
Designer
Architecture: Miao Design Studio (Design Architect), Pu Miao; Hanjia Design Group, Shanghai (Architect of Record), Jiang Ningqing
Structure: Shanghai Yuangui Structural Design Inc., Zhang Yewei, Liu Xiao
Engineering: Hanjia Design Group, Shanghai, Guo Zhong, Yu Yang, Wu Qiuyan

Publication

Green (a special volume of *Domus* Chinese Edition, October, 2011), *Time+Architecture* (4/2012).
Atlas of Contemporary Chinese Architecture (2014).

剖面		Section	
1	出租	1	Rental
2	休息	2	Resting
3	多功能外墙	3	Multi-functional wall

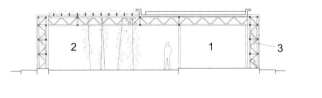

0　　　　　5m

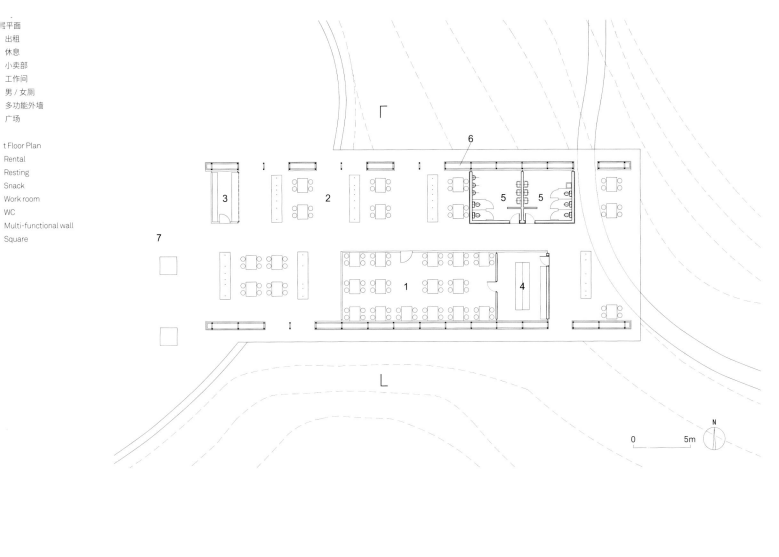

層平面
出租
休息
小卖部
工作间
男 / 女厕
多功能外墙
广场

t Floor Plan
Rental
Resting
Snack
Work room
WC
Multi-functional wall
Square

平面
屋顶
花架

Plan
Roof
Trellises

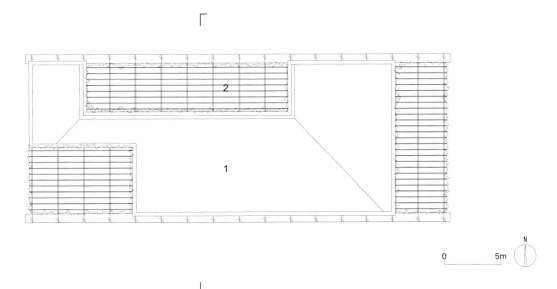

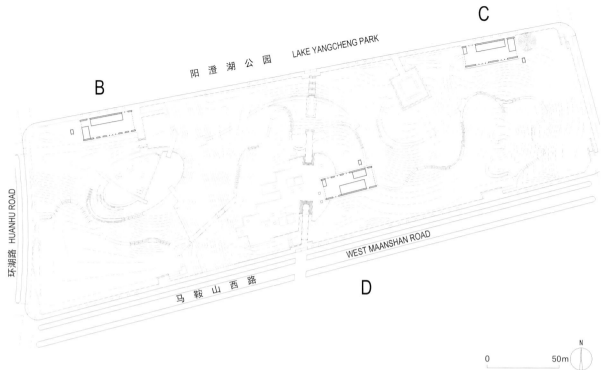

阳 澄 湖 公园 LAKE YANGCHENG PARK

B

C

D

环湖路 HUANHU ROAD

马 鞍 山 西 路

WEST MAANSHAN ROAD

0 50m

N

建筑北立面及相邻椭圆广场。
North elevation of the building and the nearby
ellipse square.

从东北角望建筑全景。
Northeast view of the building.

建筑西端开向椭圆广场，左边为小卖部。
The west end of the building opens into the
ellipse square; the snack shop is to the left.

从出租空间（室内装修尚未做）内望相邻的室外休息空间。
From the inside of rental (interior finishing not completed yet) looking into the adjacent outdoor public resting spaces.

从西南角望建筑内的室外休息空间，右面是出租空间。
The outdoor public resting space in the building seen from the southwest; rental is on the right.

从南望建筑内的室外休息空间。

he outdoor public resting space in the building,
een from the south.

北望建筑内的室外休息空间，左面为出租空间。

e outdoor public resting space in the building,
en from the north. Rental is on the left.

059

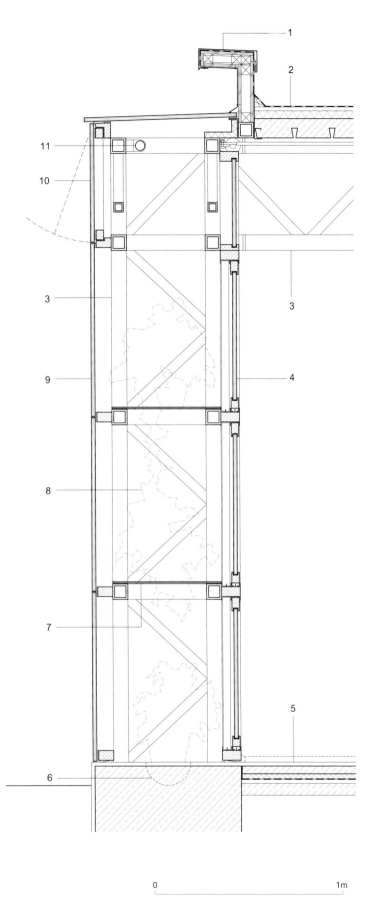

双重外墙同时包容刚架结构。
The double-skin wall contains the rigid frame.

多功能外墙剖面
1 女儿墙构造：
　　1mm 厚粉末喷涂铝板压顶
　　防水层
　　15mm 厚 OSB 板
　　钢管框架及纵向木龙骨
2 屋顶构造：
　　防水层自带保护层
　　20mm 厚水泥砂浆
　　40mm 厚硬质保温层（EPS）
　　轻质混凝土找坡层，2% 坡度，最薄处 100mm 厚
　　钢屋面板
3 刚架，80mm×80mm 钢管弦杆
4 内层幕墙（铝推拉窗）：
　　6mm 厚浮法玻璃，部分磨砂（中空面）+9mm 厚
　　空气层 + 6mm 厚钢化玻璃（室内面）
5 地面构造：
　　面层由租户提供
　　20mm 厚水泥砂浆
　　40mm 厚配筋细石混凝土
　　25mm 厚硬质保温层（EPS）
　　防水涂层
　　20mm 厚水泥砂浆
　　60mm 厚混凝土垫层
　　素土夯实
6 埋地灯

7 10mm 厚钢化玻璃搁板（北立面）或 20m
　　塑料贴面木搁板（南立面）
8 盆栽植物
9 外层幕墙（铝固定窗）：
　　5mm 厚钢化 low-E 玻璃（室外面）+ 6m
　　空气层 + 5mm 厚浮法玻璃，部分磨砂（中
10 上悬窗
11 荧光灯管

Multi-functional wall section
1 Parapet construction:
　　1mm powder-coated sheet aluminum capp
　　Waterproof membrane
　　15mm OSB sheathing
　　Steel-tubing structure with lateral wood st
2 Roof construction:
　　Waterproof membrane with protective cou
　　20mm cement mortar
　　40mm rigid insulation (EPS)
　　Min. 100mm light aggregate concrete to
　　with 2% slope
　　Steel deck
3 Rigid frame with 80X80mm steel tubing as c
4 Inner glazing in aluminum sliding-window fra
　　6mm float glass, partially sand-blasted (in
　　+ 9mm cavity + 6mm toughened glass (ou

0 _____ 1m

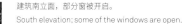

loor construction:
inished floor to be installed by tenant
0mm cement plaster
0mm reinforced concrete with fine aggregate
5mm rigid insulation (EPS)
Naterproof coating
0mm cement mortar
0mm concrete
ompacted soil
loor lighting
0mm toughened glass shelves (northern
levation) or 20mm plastic-laminated wood
helves (southern elevation)
otted plants
uter glazing in aluminum fixed-window
rames:
mm toughened glass with low-E coating
outside) + 6mm cavity + 5mm float glass,
artially sand-blasted (inside)
op-hung window
luorescent lighting tube

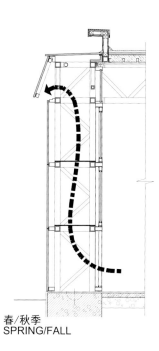

春/秋季
SPRING/FALL

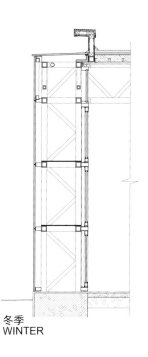

冬季
WINTER

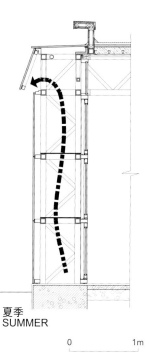

夏季
SUMMER

0   1m

多功能外墙在不同季节的节能功能。
Multi-functional wall: energy-saving in different seasons.

在出租空间（室内装修尚未做）内望南面的双外墙。

The south double-skin wall as seen from the inside of rental (interior finishing not completed yet).

从室外看双重外墙的三种通透层次。

The three kinds of depths of the double-skin wall seen from the exterior.

内看双重外墙。
double-skin wall, seen from the interior.

建筑与园林各自扮演功能配对中的
独特角色

Architecture and Landscape Each Playing a
Unique Role in a Functional Pair

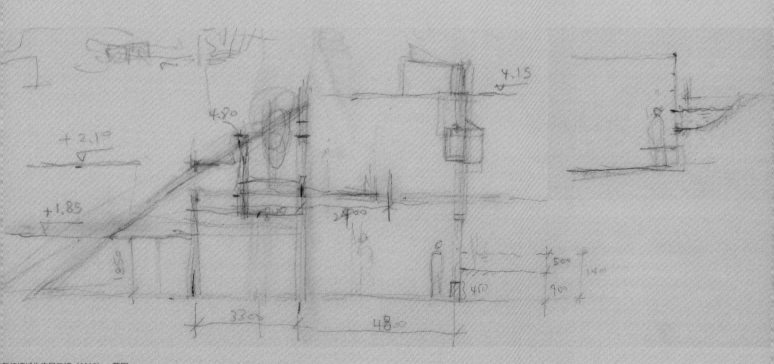

新江湾城生态展示馆（2005），草图
Jiangwan Ecological Exhibition Center,
Shanghai (2005), sketch

山中教堂
——株洲朱亭堂（2002）

A Church in the Hills
—Zhuting Church, Zhuzhou (2002)

沿湘江东岸从湖南的株洲市向南走七八十公里，在群山靠拢到江边之前的地方，有一个叫朱亭的小镇。大约十多年前，当地的一些农民自发地聚集起来做礼拜唱赞美歌。这个地方的经济虽然说通过多种经营比改革开放前要改善了好多，但由于湘江有时要发大水，农民的生活也不是完全有保障。所以当谁家遭了灾或有人生了重病，其他兄弟姐妹就组织援助并去他家为病人祷告。时间长了，参加聚会的人逐步增加到二三百人。除了本镇的人之外，还有从山那边过来的。原来用作会场的几个主持人家中的堂房就显得不够用了。在这个节骨眼上，一位过去曾下放到此地的深圳教友筹措了一笔建造费用，当地政府又给予支持，在镇南山中拿到一块约 500 平方米的地，再加上会中信徒的出钱出力，到了 1999 年初，为教徒建造一个自己的家的多年愿望终于接近实现了。我就是在这个时候，被要求为他们义务做一个建筑设计方案。任务书中的主要功能要求有一个可容纳四五百人的礼拜空间，一个供教徒聚餐团契的场所，以及几个供远处村庄走来的教徒留宿的房间。这个工程有两个"紧"：一是基地紧，二是经费紧。我的设计可以说是围绕这两个"紧"字来做文章的。

迂回空间

一个神圣空间通常需要有一个前导空间，来区分它与周围世俗空间这两个不同的领域，同时为崇拜者从后者进入前者时提供一个逐步调整心态、培养情绪的场所。这两个功能在朱亭堂中特别重要，因为基地坐落在大片森林当中，与教堂作为一个温暖的庇护所形成强烈对比。同时基地就在一条公路的边上，附近还有一个通向水口、黄龙等村镇的三叉路口，与教堂功能有闹静的区别。但由于这块呈窄长三角形的基地挤在一个山峰的侧面与公路之间的条状地带中，最宽处也不过 20 米，从公路边开始走没几步就到山坡边了，根本没有容纳大块缓冲空间的余地。

我国传统寺庙的规划原理在这里提供了启示。如果基地宽敞平整，传统寺庙建筑总是串在一条笔直的中轴线上。而山地风景区中的庙宇因地形所限，往往将香客进入建筑群的流线按等高线左右迂回。虽然在每个单体中流线还是正对建筑，但从整个建筑群来看却含有多条各种方向的轴线，只不过这是身在其中的游人不易察觉到的。

朱亭堂采用了类似因地制宜的做法，索性将教堂（含礼拜堂及辅助建筑）的长向沿窄长基地的长向布置，将教堂与公路之间的剩余地带做成一条与教堂平行的前导空间，利用迂回人的流线的做法在浅进深的基地中取得纵深感。由于信徒大部分来自北面山下的朱亭镇，当他们攀登到离教堂还有一段距离时，在一个高高的十字架从教堂围墙后面伸出之处，信徒就可以进入一条沿围墙外延伸但隐蔽在竹篱后的小径。在二十余米的小径尽端，人们将跨越一座石板桥（桥下是两座山峰之间的集水坑，供当地农民取水用）。

由于建筑必须采取规则的矩形平面（见下文），到石板桥处公路已逼到房脚下，所以过桥后人们必须先暂时进入建筑内一个类似门厅的三角形空间，一道室内斜□将该空间与礼拜堂分隔开。但人的流线马上一拐又回到室外，进入一连串带绿化石凳的小院。在礼拜开始前各村来的信徒可以在这里社交，这也是教会活动的一□重要组成部分。小型的院落不仅更具亲切感，同时也为未来出现的高大礼拜堂开□反衬。这一空间序列终止在一个较大的礼拜堂前院中，从这里人们再一转身，就□对二层高礼拜堂的正门了。

可望而不可即

如何用建筑手段来表达崇拜对象的神圣感呢？宗教史学家米尔恰·伊里亚□（Mircea Eliade）对"神圣"的本质有一个独到的见解："值得注意的是人对所□神圣事物（广义上来说）的矛盾态度。一方面他希望通过接触神迹来保护与强化自□另一方面他又不愿被提升到一个高于自然和世俗的层面，从而完全失去自身。他□望超越自身但又不愿彻底离开它。"这种模棱两可的态度决定了神圣空间必然同□是"可进入和不可进入的"[1]。用建筑语言来说就是可望而不可即的。

但在目前一般的教堂设计中，圣坛通常是一堵挂有十字架的山墙，较好的也□过是在墙上开一些彩色玻璃窗，或在十字架处设计某种特殊的光影效果，在空间□始终给人一种明确的终点站之感，而这恰恰不是"神圣"的特点之一。事实上，□史上最优秀的教堂设计案例已经意识到这个问题，如哥特式天主教堂中常用透□铁栅门及唱诗班空间将圣坛与教徒座席远远隔开，使前者在后者眼中遥不可测。□巴黎的圣洛克（St.Roch）教堂里，当我走到圣坛的十字架时却发现它并不是终点，□它后面的墙上开有洞口，透过洞口又可以看到另外几个洞口，在其最深远处隐□示出一尊基督受难的雕像，从而在实际的圣坛后面又创造了一个虚拟圣坛。

朱亭堂不可能采用上述这类昂贵的手法。但我国的建筑传统提示了另一个□不多却有类似效果的处理方式。如果说西方建筑倾向于将室内外空间处理成两□相独立的系统（像教堂大多是一个与室外隔绝的内向空间），中国人从来喜欢□个房间与一个室外空间配成对，共同服务于某个生活功能。这种环境结构在今天□很受欢迎。除了在前面说的社交院落中应用了这一概念外，我们还在这里试图用□它来创造神圣空间。在窄长三角形基地上，较宽部分布置了教堂后，其北面还剩□一段三角形的尖端空间，这一森林环绕的室外空间正好可以服务于我们的目的。

当信徒从庭院进入礼拜堂时，在传统的圣坛处他们会看到一个 6.6 米高的"门洞□他们脚下的走道穿过封闭门洞的大玻璃，一直向室外的远方伸去，爬上基地北□满森林的山坡，从那里一个遥远的十字架在树梢上的天空中显露出来。如果从□的围墙外看，这里恰恰是访问者第一次看到十字架上部、进入前导空间之处。从□

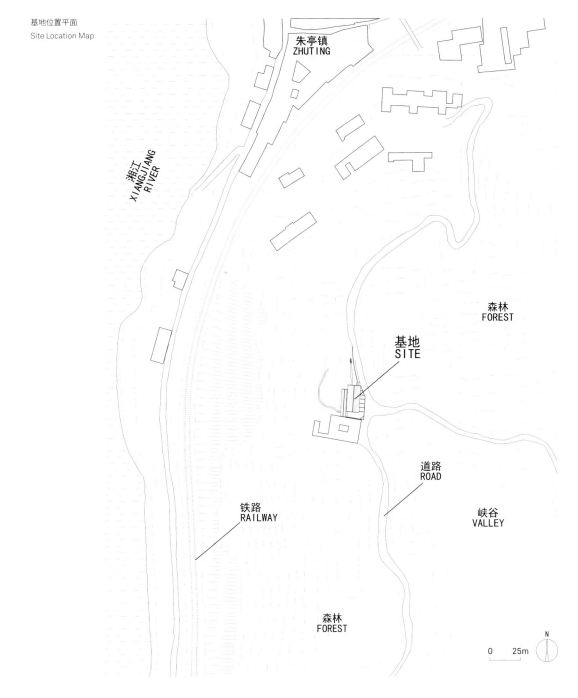

基地位置平面
Site Location Map

朱亭镇
ZHUTING

湘江
XIANGJIANG
RIVER

森林
FOREST

基地
SITE

道路
ROAD

铁路
RAILWAY

峡谷
VALLEY

森林
FOREST

0   25m

N

他们来到"门洞"处时几乎走完了一个 U 形路线，终于看到十字架的全景。"门洞"
面有一面斜墙，它一方面将上面说过的门厅与礼拜空间分隔开来，另一方面成为
布讲坛的背景，使他能在"门洞"边为教友们引路。

地参与

教堂与独家住宅的设计有某种相似之处，使用者对建筑都有很强烈的个人感情联
因为信徒把教堂看成自己的第二个家。在朱亭堂的建设中，教会成员参与了从总
布局到细节设计、从材料选择到施工的全过程。这种参与在一定程度上防止了建筑
召搬自己熟悉的（西方）建筑语言，促进了对当地形式与材料的发掘与创造性利用。
不仅保证了将成本控制在十四五万元内，更打开了我们对现代建筑本土化的视野。
比如说，建筑的基本形式坚持用最简单的，也是本地民居常用的矩形平面上加

双坡屋顶，因为随便扭曲一下建筑造型就会突破成本。独特的空间效果是通过在不
破坏规则的总体布局的前提下做局部调整来产生的。礼拜堂内利用靠近地面的低窗
及离地 5 米多的高窗组织对流通风。在结构形式上，采用当地施工队熟悉的钢木屋架、
砖承重墙及砖墩内框架体系。利用类似飞扶壁的室外横墙来加强结构的侧向稳定，
这些横墙同时又起到分隔社交小院、制造纵深效果的作用。在建筑材料上，选择了
最便宜的本地烧制的红砖红瓦（礼拜堂的屋顶仅为冷摊瓦）及木门窗。室外除钟楼
及围墙外不做粉刷，利用清水砖瓦本身的窑变色调、质感及绿化，来丰富建筑立面（现
在看来连上述的部分粉刷都是多余的）。室内大部分也仅粉刷了一道石灰水。

大概花了两年时间，朱亭堂终于在克服了各种想不到的困难后基本建成了。用
目前的建筑时尚来衡量，这座总建筑面积只有约 380 平方米的乡村教堂实在是既小又

土气。由于部分施工是由教友自己完成的，不少细节看上去还有点粗糙。但我觉得它至少很真实，恐怕比许多大城市的形象工程更真实地反映了自己的地点、时代和使用者。这同时使我联想到童年初次走进历史建筑时的惊奇发现，一方面它们都明显地要遵循某个理想原型，但另一方面，从总体布局到地面铺砌，我经常会发现微小的不对称或不完全一致的重复。而正是这种"粗糙"，使这些古迹具有后来大量出现的"准确"的假古董所没有的生命力。

我的体验后来在听克里斯朵夫·亚历山大（Christopher Alexander）的讲课时得到证实。他甚至把粗糙（Roughness）看成是他的十五条产生完美几何结构的原则之一。他指出："真实存在的东西必须适应于不规则的外部环境，它们因此变成不规则的了。"亚历山大认为圆满的构图之所以必然会有少许不准确的地方，其根本原因在于设计者要能集中注意力在基本的大关系上，就必须在细节上允许微差或局部妥协 [2]。这一原则在工业化社会中的运用恐怕还有待探讨。但在朱亭堂中，要在狭窄的基地中实现上述宏观的几何构图及让使用者参与建设过程，就必须接受局部的不规则和不准确。事实上，这些外在条件的约束往往意外地产生积极效果。上面提到的建筑入口处的集水坑，迫使我们增加了一座小桥，但桥作为一个符号反而强调了进入神圣领域的意义。基地的局限使来访人流必须穿越建筑一角再回到室外，结果却丰富了原本简单的矩形平面。我不敢说最后的结果达到了圆满，但希望它能揭示一个中国内地乡村教堂的本质。

现在每到圣诞节，周围远到三十多里地以外的信徒都会赶来聚会。这次去拍照时发现，他们在基地西南角的山坡上又加建了厨房（二期），从一期室内楼梯的中间平台进入。这使朱亭堂更有家的感觉。如果从厨房的后门出去再向上攀登十几步，可以爬到山顶的树林中，山那边望下去就是湘江了。（2007）

注释

[1] Mircea Eliade, *Patterns in Comparative Religion* (New York: The New American Library Inc., 1963), pp.17-18, 384.

[2] Christopher Alexander, *The Nature of Order: An Essay on the Art of Building and the Nature of the Universe* (未出版的早期手稿, 1986), pp.292-293。该书的正式版本已出版 (Berkeley, CA: Center for Environmental Structure, 2003-4)。

Abstract

Situated in the mountains next to a small town Zhuting, the church provides the first permanent place of worship for the congregation of 200-300 surrounding peasants. The volunteered design focuses on a tight site and a tight budget, aiming to localize Modernism in the rural area of a developing country.

Squeezed between the hillside and a dirt road, the narrow site makes it difficult to provide a transitional space from the profane to the sacred. Learning from the traditional Chinese temples on hills, this design creates a zigzag spatial sequence to produce the needed prologue. People go through a path shielded from the road by a bamboo fence, pass a string of five small social courtyards, and arrive at a forecourt where they turn around and, only then, enter the worship space.

The design also tries to create a unique interpretation of the sacred as "visible but not reachable." Instead of using expensive lighting effect or ornaments, the design moves the cross from the interior into the pristine forest outside, which is inspired by a Chinese architectural tradition that pairs indoor and outdoor spaces together to serve each function. Upon entering the church, people will see the central aisle goes through a 6.6-meter high "gate" into the forest where a distant cross hovers above the trees, completing a U-shaped journey that starts under the same cross.

The design adopts the form and material of local farm buildings to reduce costs and to facilitate users' participation in the design and construction process.

工程资料

地点 湖南省株洲县朱亭镇朱亭林场
时间 1999—2002
基地面积 500 平方米
建筑面积 380 平方米
业主 朱亭堂
设计
建筑：缪朴、湖南省株洲县建筑设计院
配合工种：湖南省株洲县建筑设计院

发表

《时代建筑》（2007 年第 4 期）。

Project Data

Location Zhuting, Zhuzhou, Hunan Province, China
Project Period 1999-2002
Site Area 500 square meters
Floor Area 380 square meters
Client Zhuting Church
Designer
Architecture: Pu Miao (Design Architect), Zhuzhou County Architectural Design Institute, Hunan Prov (Architect of Record)
Consultants: Zhuzhou County Architectural Design Institute, Hunan Province

Publication

Time+Architecture (4/2007).

（上）东立面，（下）剖面　　(Upper) East Elevation, (lower) Section

1	礼拜堂	1	Church
2	宿舍	2	Dormitory
3	社交空间	3	Communal space
4	庭院	4	Courtyard

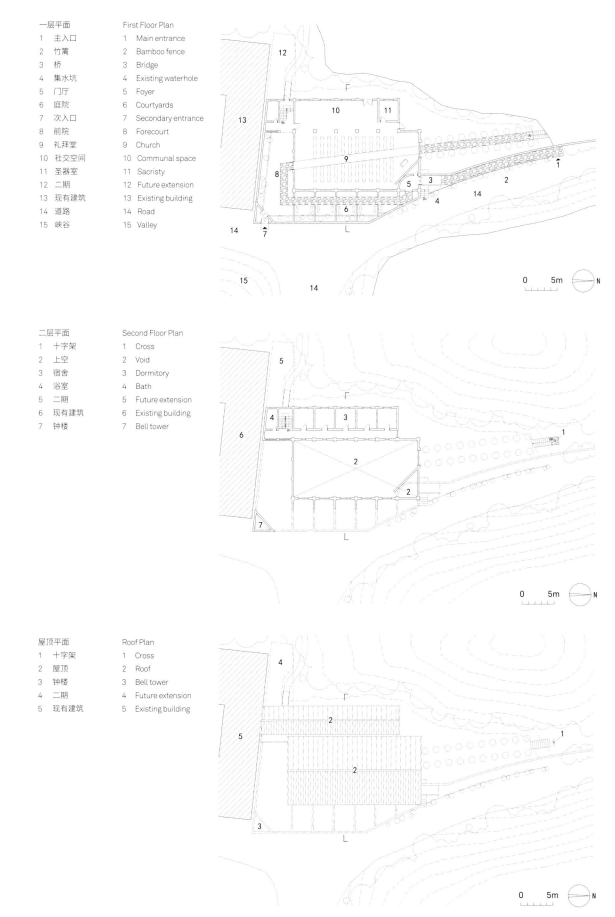

一层平面　First Floor Plan
1　主入口　1　Main entrance
2　竹篱　2　Bamboo fence
3　桥　3　Bridge
4　集水坑　4　Existing waterhole
5　门厅　5　Foyer
6　庭院　6　Courtyards
7　次入口　7　Secondary entrance
8　前院　8　Forecourt
9　礼拜堂　9　Church
10　社交空间　10　Communal space
11　圣器室　11　Sacristy
12　二期　12　Future extension
13　现有建筑　13　Existing building
14　道路　14　Road
15　峡谷　15　Valley

二层平面　Second Floor Plan
1　十字架　1　Cross
2　上空　2　Void
3　宿舍　3　Dormitory
4　浴室　4　Bath
5　二期　5　Future extension
6　现有建筑　6　Existing building
7　钟楼　7　Bell tower

屋顶平面　Roof Plan
1　十字架　1　Cross
2　屋顶　2　Roof
3　钟楼　3　Bell tower
4　二期　4　Future extension
5　现有建筑　5　Existing building

竹篱与围墙之间的前导小径，远处为教堂入口。
Between the bamboo fence and the wall, the entering path leads to the main entrance.

前导小径尽端的石板桥及教堂入口。
A bridge before the main entrance; under the bridge is a preserved waterhole.

从北向南望社交小院序列及钟楼。
A north-south view of the series of courtyards and the bell tower.

在门厅中仰看分隔礼拜空间与门厅的斜墙。
The interior of the triangular foyer; the tall diagonal wall separates the foyer from the worship space.

从南向北望社交小院序列。
A south-north view of the series of courtyards.

社交小院中的绿化。
One of the courtyards designed for social activities.

及教堂东面外景。
ell tower and the east exterior of the church.

通往室外圣坛的"门洞"及右侧斜墙。
The "gate" leading toward the exterior cross and the diagonal wall separating the worship space from the foyer.

通往室外圣坛的"门洞"。
The "gate" leading toward the exterior cross.

外十字架处回望教堂，左侧围墙外为前导小径。
n the cross looking back toward the church;
ide of the left wall is the entering path.

留宿房间内景。
Interior of a lodging room.

寓居佛性的园林
——长兴寿圣寺扩建（方案，2008）

Landscape as Where the Sacred Resides
—Expansion of Shousheng Temple, Changxing (Scheme, 2008)

群山竹林环抱中的寿圣寺始建于公元 3 世纪，其后经几度兴废。1990 年代开始重建，到 2008 年已经形成了含宝塔的庙堂区及僧人生活区的建成部分（见寿圣寺总体规划平面）。寺庙另外还拥有几块可供开发的地皮，如紧邻建成区东边的竹林地及飞地（与竹林地之间有村路分隔）。方丈有意在这两块空地上建设一个禅修中心发挥寿圣寺位于顾渚风景区中的优势，定期为周边城市居民举办禅学夏令营、研班等活动。我有幸被邀请为此扩建计划提供一个初步的规划与建筑设想。

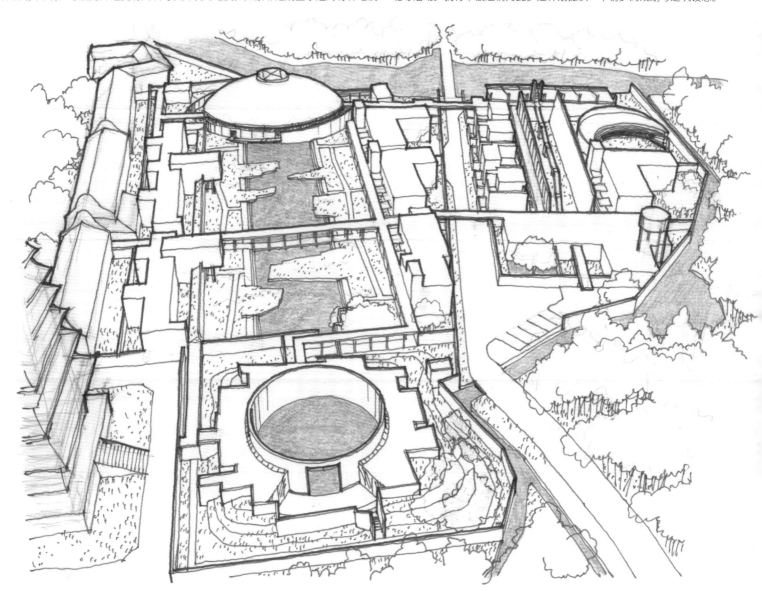

图 1. 禅修中心鸟瞰（基地内大部分树木未画）。
Figure 1. An aerial view of Buddhist Study Center (most trees in the site are omitted).

划分区

为了确保扩建后的新旧两部分能完美结合，特为整个寺庙做了一个新的总体规划。在新规划中，庙堂区（含塔院及对外餐厅）对所有人开放。夹在庙堂区及禅修中心北部之间的僧人生活区仅为寺庙内部人员使用。庙堂区东面的禅修中心为学员及夜访客服务，它分居竹林地及飞地上的两部分通过南北两座天桥连为一体。禅修中心可从飞地门厅或寺内餐厅处的两个入口进入，两个入口之间由一条沿东西向横贯整个中心的敞廊连接。新规划并按以上分区构想调整了建成部分的围墙等。

本策略：有宗教意义的环境

新建的禅修中心不能就是一个按物质功能组织的青年中心或旅游旅馆；它必须造出有宗教象征意义的环境或"景观"，用自己的空间来帮助学员体会禅义。国内许多新建寺庙习惯用符号式的庸俗象征来达到这一目的，例如到处布置仿古建筑形象或布置宗教偶像。与此不同，本禅修中心方案采用一种现代建筑风格。它利用原就必需的建筑空间或部件（如庭院、消防水池、坡道、天窗等），适当调整其平面布局、几何造型及用材，使其在满足物质功能的同时又能形成宗教象征。

在运用这一基本策略时，本方案特别强调了用园林景观（像水体、光影、绿化等）象征神圣，取代具象的偶像。相信这应更接近禅宗"无念为宗，无相为体"的教义，用自然元素演示佛性，也延伸了南北朝名僧竺道生有关"一切众生，皆有佛性"的理论以及他在虎丘说法、顽石为之点头的传说。与建成区用钢筋混凝土模仿古典式的做法相比，这种较抽象的象征更适合于禅修中心的主要服务对象——中青年城市居民。

本策略的具体应用

禅修中心总体：两个中心庭院，三种形式语言

本规划在竹林地及飞地各布置了一个中心庭院，不仅满足了组织各地块内建筑群的交通等功能需要，同时通过建筑与园林相结合，形成富有宗教联想的中心景观（图1）。

在竹林地中是一个贯穿地块南北的大水池。水在佛教中自古就有净化心灵、引向彼界的意蕴。在基地中央廊桥上的访问者可看到水池向南北延伸，将视线引领到位于基地两端的 200 人禅堂及陈列馆（兼为圆成法师纪念堂），点出了两者较高层次

的宗教意义。水池的两边还设计了多个上植竹林的半岛插入水中。当从廊桥向 200 人禅堂及陈列馆望去时，两边的层层竹林如同舞台上的侧幕，使远处终端建筑显得遥不可及而更觉神圣（图 2）。学员住宿设施分列水池东西两边，既表示从属的性质，也方便了住宿者到池边观赏水景领会禅意。为此水池的大部分边缘设计成让人触碰水的台阶式。该水池除了满足防火要求外，也为建筑群奠定了富有宗教含义的布局。

将飞地中各单体组织起来的中心景观是一个内含坡道的庭院（图 3）。设置此坡道是由于飞地与寺庙主体只能通过天桥连接，必须为此提供无障碍通道。但本设计中的坡道已超越了功能。它位于一个由两侧竹林及高耸斜墙（经文墙）形成的"峡谷"中，清水混凝土斜墙上满书佛教经文，使交通与修习两者合一。坡道爬升到天桥高度（4.5 米）后再向上延伸 1.5 米，来到一个可远眺北面远山及天空的瞭望平台。以上整个布置让使用者感受到从尘世接近彼岸的艰苦历程以及最终为空的境界（图 4）。

图 3. 经文墙庭院。
Figure 3. The courtyard of scripture wall.

沿水池北望 200 人禅堂。
Figure 2. Looking north along the pool toward 200-person meditation hall.

图 4. 经文墙庭院坡道北端的瞭望平台。
Figure 4. The lookout platform at the north end of the ramp in the courtyard of scripture wall.

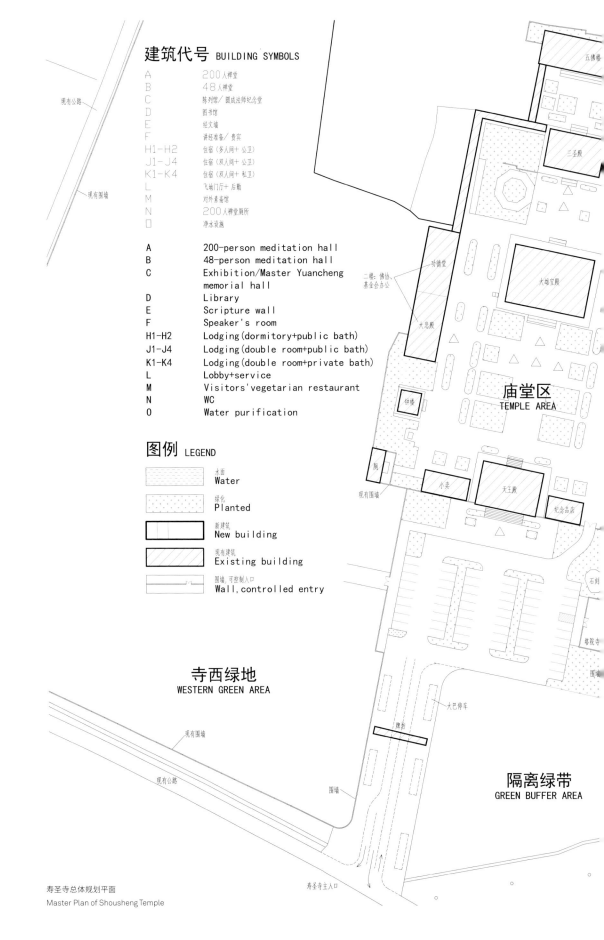

建筑代号 BUILDING SYMBOLS

A	200人禅堂
B	48人禅堂
C	陈列馆／圆成法师纪念堂
D	图书馆
E	经文墙
F	讲经准备／贵宾
H1-H2	住宿（多人间＋公卫）
J1-J4	住宿（双人间＋公卫）
K1-K4	住宿（双人间＋私卫）
L	飞地门厅＋后勤
M	对外素斋馆
N	200人禅堂厕所
O	净水设施

A	200-person meditation hall
B	48-person meditation hall
C	Exhibition/Master Yuancheng memorial hall
D	Library
E	Scripture wall
F	Speaker's room
H1-H2	Lodging(dormitory+public bath)
J1-J4	Lodging(double room+public bath)
K1-K4	Lodging(double room+private bath)
L	Lobby+service
M	Visitors'vegetarian restaurant
N	WC
O	Water purification

图例 LEGEND

水面
Water

绿化
Planted

新建筑
New building

现有建筑
Existing building

围墙，可控制入口
Wall,controlled entry

现有公路

现有围墙

五佛楼

三圣殿

大雄宝殿

二楼：佛协、
基金会办公

功德堂

大悲殿

庙堂区
TEMPLE AREA

钟楼

厕

小卖

天王殿

纪念品店

现有围墙

石刻

塔院寺

围

寺西绿地
WESTERN GREEN AREA

现有围墙

现有公路

围墙

大巴停车

摊挡

隔离绿带
GREEN BUFFER AREA

寿圣寺主入口

寿圣寺总体规划平面
Master Plan of Shousheng Temple

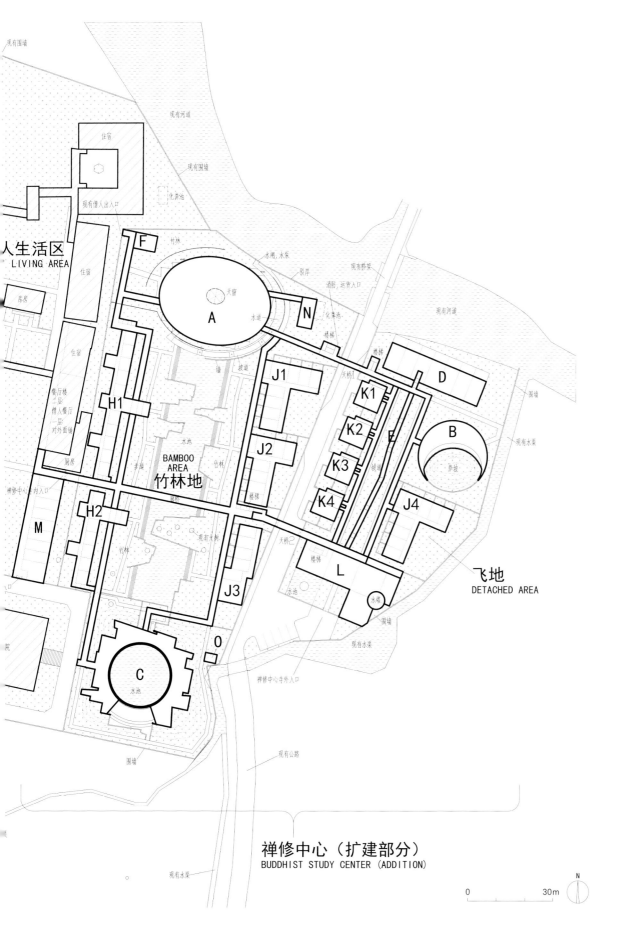

人生活区
LIVING AREA

F

A

N

H1

J1

D

K1

BAMBOO
AREA
竹林地

J2

K2

E

B

K3

J4

M

H2

K4

飞地
DETACHED AREA

J3

L

O

C

禅修中心（扩建部分）
BUDDHIST STUDY CENTER (ADDITION)

0   30m

N

沿斜墙墙脚外侧为一圈带休息座位的有顶敞廊，连接飞地中各建筑单体，同时也是学员之间交流修习心得的地方。斜墙在敞廊处开有多个细长景窗，可一瞥经文墙庭院中的竹林。斜墙外侧立面满布爬山虎绿化。

为了加强上述用空间布局创造出来的景观的感染力，本规划同时为环绕两个中心庭院的建筑实体设定了三种形式语言，各自也含有不同的宗教意义。

圆形——圆在佛教中象征领悟真理的最高境界。新建部分中的三个团体修习场所——200 人及 48 人禅堂以及陈列馆均采用圆或椭圆造型。

折线形——它给人以不规则之感，同时也有对自然岩石的联想。在椭圆形 200 人禅堂的周边硬地中，大水池的两岸半岛以及陈列馆的周边轮廓上使用了折线形式，寓意尘世"无明"人生的艰坎坷，作为目标地的圆形空间相形之下给人以安慰纾解之感。引入不规则构图也有利于保留竹林地中近十棵现有金钱松。

方形——该中性的形式用于所有服务物质功能的建筑单体，为上述两种形式提供了一个较为低调的背景。

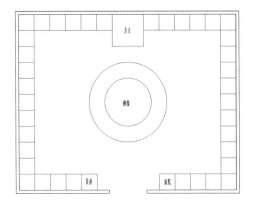

图 5. 传统禅堂平面。
Figure 5. Plan of a traditional meditation hall.

2. 200 人禅堂

禅堂是教徒"凝心于内，令得静止，成就一向，而获禅定"的场所。打禅者均坐在地面的座垫上，所以其空间应强调亲切平静，不能过分高耸而使使用者感到被疏离。主要景观还应接近坐下的人眼高度。传统的禅堂平面大多沿外墙布置一圈座位（图 5）。显然，该平面不能满足要求的 200 人容量，因此必须做成多排座位，但多排座位又会遮挡后排看讲道法师的视线。

结合总体布局设定的圆形及上述功能需要，本建筑设计创造了一个椭圆盆地形

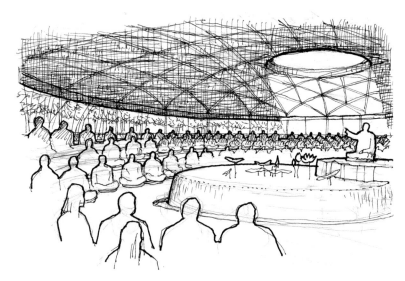

图 6. 200 人禅堂内景。
Figure 6. Interior of 200-person meditation hall.

空间，其中心为一莲花池，旁设"巨石"状混凝土讲经坛。莲花池水面高出禅堂地面米，池内有一涌泉，讲道时池水会沿池壁四周潺潺流淌到地面的水渠中。在莲花内放置盆栽莲花，用自然景物代替传统的人造偶像，同时又不会遮挡信徒看到讲道正对莲花池及讲经坛上方为一天窗，打下一缕阳光（图 6）。

莲花池及讲经坛四周设一圈 2~3 米宽的平地供跑香用。（该仪式还可以考虑在堂周围屋檐下硬地上进行，或者甚至延伸到大水池周边，使跑香行列穿行在竹木之间。）围绕上述中心区域为台阶型打坐空间，保证了每排人看讲道者及莲花池面的视线不受干扰。以上盆型空间创造了一种重心向下的安定感，有助于在参修的学员、法师及佛的象征之间建立起一种凝聚感。本设计也考虑了禅堂转化为场使用的可能。

禅堂周边全为玻璃墙，室外依次为一圈屋檐下的硬地、一圈竹林及一道磨璃围屏。从室内望出去，整个禅堂 84 米长的周边是一道透明边界，参禅者好像被竹林环抱。禅堂屋顶为一椭圆球形屋盖，采用钢管网壳结构，从室内可以看禅堂主立面（南立面）被多重"表皮"包裹，所显示的半隐半显之感强调了圣俗立面西半部的"表皮"是上述磨砂玻璃围屏，建筑东半部被从飞地过来的坡道所掩盖。只有从立面中央可以一瞥禅堂内被顶光照亮的莲花池（图 2）。

3. 48 人禅堂

此禅堂类似园林中的亭榭，室内与室外不完全隔断，是一个适合小团体做转意的讲道打坐或其他类似活动（如禅茶处）的场所。本建筑设计为一个坐北朝

图 7. 绿草在背光下的实
Figure 7. Backlit grasse

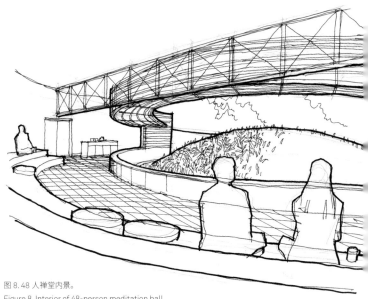

图 8. 48 人禅堂内景。
Figure 8. Interior of 48-person meditation hall.

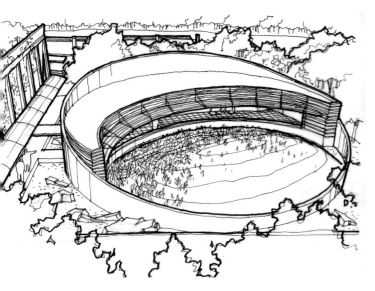

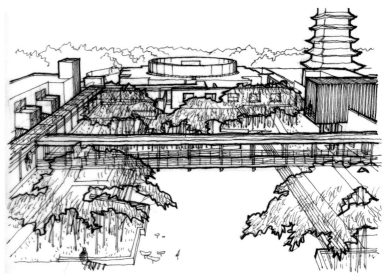

图 9. 48 人禅堂环抱草坡。

Figure 9. The 48-person meditation hall embraces a grassy mound.

图 10. 沿水池南望陈列馆。

Figure 10. Looking south along the pool toward exhibition hall.

形建筑，环抱一个满种青草的椭圆形土丘。使用者坐在建筑内台阶型座凳上，细品味背面被南来阳光照亮的萋萋绿草及其象征的寓于众生之中的佛性（图 7、9）。这一景观能随季节转换而不断变化，体现了"无常"的概念。

由于采用了大跨度钢结构（由一榀直线主桁架及两榀曲线次桁架组成），该景观超过 18 米，其间不受任何柱子等构件的干扰。整个建筑及土坡被一道圆形斜顶混墙环绕，墙高从土坡处的 0.9 米升到建筑处的 5 米余。圆墙外的庭院地面铺砌切不规则立体形状的当地石材，与墙内绿草形成对比，暗喻充满烦恼的尘世（图 9）。外立面为清水混凝土及本色木格栅（内衬玻璃），以强调接近自然之意。

陈列馆（兼为圆成法师纪念堂）

作为大水池南端的标志建筑，本单体的功能为陈列业主收集的佛教文物及纪念寺发展中有重大贡献的僧人。本建筑外观好像一座小山，中心是一圆桶形高墙（纪）。周围环绕不规则折线形的展览空间，展览空间外为一圈逐步低下的绿化土坡。00 人禅堂面向大水池的南立面相呼应，本建筑的北立面也被两重"表皮"所遮掩，是从西面延伸过来的住宿（H2）栋走廊外墙（墙上开洞口），以及从东面住宿（J3）过来的花架玻璃顶走廊。大水池在上两者之间的豁口处变成涓涓小溪，流进建（图 10）。

小溪流入室内后一分为二，环绕圆形高墙流向建筑后部。高墙上将镶嵌有关寿圣史上重要僧人（如圆成法师）事迹的石碑。陈列馆屋顶与圆形墙之间留有一圈天窗，上的小溪将观众与碑墙分开，这两个手法均突显了圣俗两界有别之意。与中心的墙相对，室内四周为折线形外墙，为各式展品提供了多种空间（图 11）。

当参观者来到陈列馆后部时，将发现一个内设多层台阶型座凳的扇形庭院，面对高墙上的一个巨大门洞。参观者从门洞中将看到圆形高墙内其实没有其他建筑，是一个简单的充满水的空间。除了光影随着时间推移变化外，一切趋于宁静虚无。用建筑语言揭示了"一切诸法，本体空寂"的佛教基本教义（图 12）。

图书馆、经文墙

经文墙庭院详见前文关于"两个中心庭院"的介绍。

与高耸的经文墙形成对比，图书馆水平展开在墙北端之下（图 13）。与图书馆北行流淌着一条现有小河，随季节变化会展示从潺潺流水到汹涌波涛等各种河景，

图 11. 陈列馆内景。

Figure 11. Interior of exhibition hall.

图 12. 从扇形庭院望中心圆形水庭。

Figure 12. From the fan-shaped courtyard looking into the central water courtyard.

它的自然砂石岸与河对岸的大片竹林更为河景增添了优美画框。而河的流转不息、易逝难追常被佛教用来演绎人生"无常"的概念。为了充分利用这一富含象征意义的现成景观，特将图书馆沿河立面设计为一道无框玻璃幕墙，并采用钢索悬挑屋盖结构体系，免除了使用柱和承重墙会对河景产生的遮挡，从而为读者展现出一幅 35 米长、2.78 米高的"山水长卷"（图 14）。

6. 住宿

以下以多人间配公用卫生间的住宿设施 (H1、H2 栋) 为例。该类型主要面向宗教夏令营、研修班等的学员，以未婚的学生或青年白领等为主。为节约造价及用地，本设计没有采用传统的旅馆房间布置，而是用 2.1 米高的轻质隔断形成个人小间。每间仅约 2.2 米见方，内设固定床榻一张，外加小书桌、椅子、挂衣设施各一。

图 15. 住宿（多人间 + 公中的个人小间内景。
Figure 15. Interior of a personal compartment lodging (dormitory+pub bath).

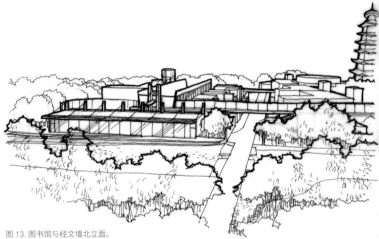

图 13. 图书馆与经文墙北立面。
Figure 13. The north elevation of library and scripture wall.

图 14. 图书馆内景。
Figure 14. Interior of library.

但就是在这样狭小的空间中，本设计仍然没有忘记创造有宗教象征意义的环境。考虑到部分学员会有在自己的私密空间中打坐的需要，故将可能兼为打坐用的床榻全长靠窗（一层的住宿者还可在窗外木平台上打坐）。窗外设计了一个袖珍庭院，虽只有 2.1 米深，但种有纤巧的绿化。按照佛典"一花一世界"的说法，这个袖珍庭院可成为打坐者凝视、参悟的对象（图 15）。二层的小间窗外，为同样目的，在邻近的敞廊屋顶上设计了屋顶绿化。(2008)

Abstract

Established in the third century, the ruined Shousheng Temple rebuilt its tem area (including the pagoda) and monks' living area by 2008 (see the Master Pl The rural temple then planned to use two adjacent vacant lots, the bamboo a and the detached area, to build a new Buddhist Study Center for summer car serving residents of nearby cities.

The Study Center is entered from the temple and the detached areas, with the entrances connected by an east-west veranda. Elevated footbridges unify the parts of the Study Center in the bamboo and the detached areas.

As the basic design strategy, the scheme goes beyond merely satisfying mate functions. It produces an environment rich in religious symbolism, using its spa to help students to understand the teaching. Instead of imitating traditic temple architecture or displaying religious idols, however, the symbolism is cre through a Modernist approach that uses functional elements (such as garden, f control reservoir, ramp, and skylight) to convey religious meanings. In particu landscape features like water, sunlight, and plants are designed to represent sacred, echoing the Buddhist teaching that "all living beings have Buddha natur

Looking along the south-north long pool (also for fire-control) that links the two m buildings in the bamboo area, the flowing water and the layers of bamboo gro flanking the pool heighten the remoteness and sacred status of the two destinatio

Inside the 200-person meditation hall, a lotus pond with a fountain replaces conventional statues. Surrounded by stepped meditation seats, the pond reinfo the calmness of a meditation place. In the exhibition hall, water from the long feeds into the central void encircled by the building, illustrating a sense of emptine

Amid the detached area, a ramp rises between two walls inscribed with Budd scripture. It not only provides access to the elevated footbridge, but also impli hard journey to enlightenment. At the top of the ramp people can enter a cantilev lookout platform, seeing empty sky and elusive mountains in the distance.

Plants play an important role in creating symbolic landscape. In the 48-per meditation hall, people contemplate the grass, backlit by the sun, on a sm mound. Even in the dormitory, students can meditate while gazing at greenerie a small private courtyard or a roof planter.

资料
浙江省长兴县水口乡
2008
面积 7 302 平方米
浙江省长兴县寿圣寺
缪朴设计工作室（缪朴）

ect Data
ation Shuikou Xiang, Changxing County,
iang Province, China
ect Period 2008
r Area 7,302 square meters
nt Shousheng Temple, Changxing County,
iang Province
gner Miao Design Studio, Pu Miao

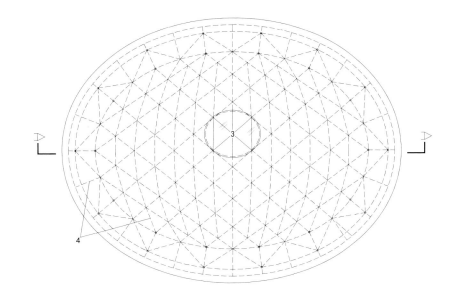

） 200 人禅堂屋顶平面，（下）A-A 剖面
莲花池、涌泉
讲经坛 / 讲台
天窗
屋顶网壳构架（室内面）

er) 200-person meditation hall, Roof Plan,
er) A-A Section
otus pond with a fountain
Podium
Skylight
Roof gridshell frame (interior side)

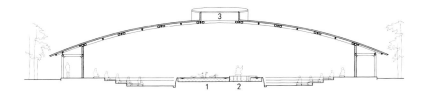

人禅堂一层平面
莲花池、涌泉
讲经坛 / 讲台
主席台（开会用）
上为天窗
跑香空间
敞廊屋檐下硬地
竹林
碧砂玻璃围屏
水道
水池

person meditation hall,
Floor Plan
otus pond with a fountain
Podium
Stage (for conferences)
Skylight above
Space for walking meditation
Paved area under veranda roof
Bamboo grove
Frosted glass screen
Water channel
Pool

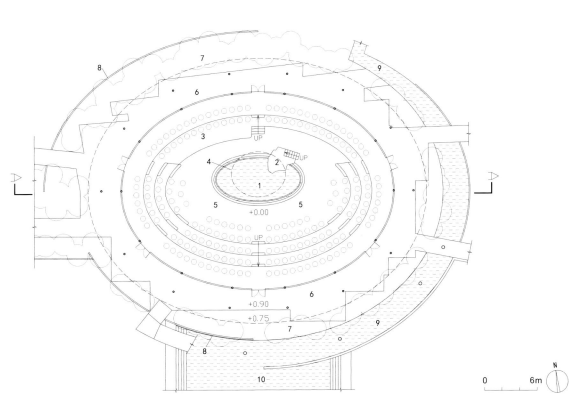

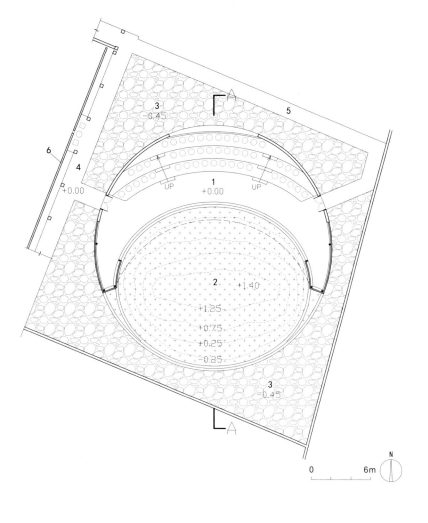

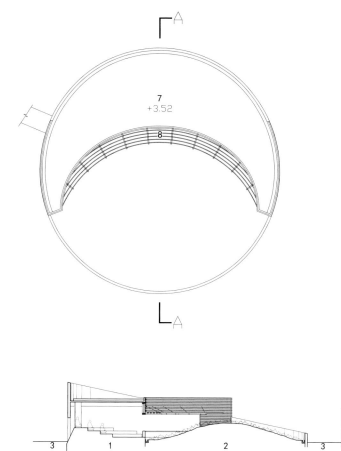

0	6m

N

（左）48 人禅堂一层平面，（右上）
屋顶平面，（右下）A-A 剖面

1　禅堂
2　草坡
3　碎石
4　敞廊
5　步道
6　经文墙（E）
7　屋顶
8　遮阳格栅

(Left) 48-person meditation hall, First Floor Plan, (right
upper) Roof Plan, (right lower) A-A Section

1　Meditation hall
2　Grassy mound
3　Gravel
4　Veranda
5　Footpath
6　Scripture wall (E)
7　Roof
8　Sun-shading trellis

（右下）陈列馆 A-A 剖
面，（右上）B-B 剖面

1　水庭
2　水渠
3　纪念墙
4　陈列
5　天窗
6　敞廊
7　扇形庭院
8　山坡
9　步道
10　塔院
11　水池

(Right lower) Exhibition hall
Section, (right upper) B-B Sec

1　Water courtyard
2　Water channel
3　Memorial wall
4　Exhibition
5　Skylight
6　Veranda
7　Fan-shaped courtyard
8　Mound
9　Footpath
10　Courtyard of the pagoda
11　Pool

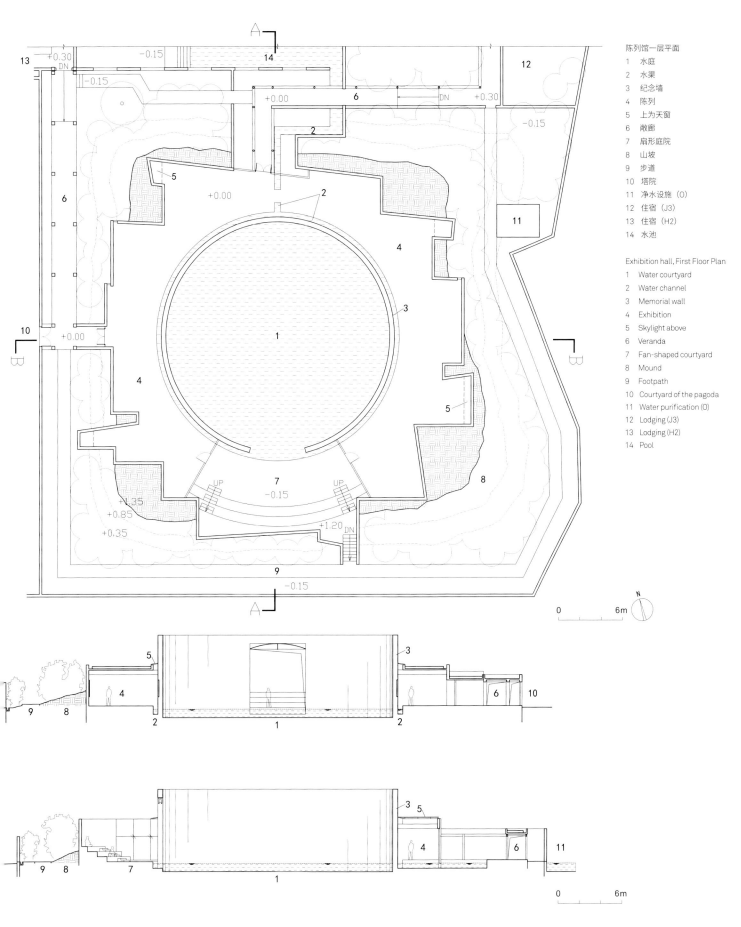

陈列馆一层平面

1 水庭
2 水渠
3 纪念墙
4 陈列
5 上为天窗
6 敞廊
7 扇形庭院
8 山坡
9 步道
10 塔院
11 净水设施（O）
12 住宿（J3）
13 住宿（H2）
14 水池

Exhibition hall, First Floor Plan

1 Water courtyard
2 Water channel
3 Memorial wall
4 Exhibition
5 Skylight above
6 Veranda
7 Fan-shaped courtyard
8 Mound
9 Footpath
10 Courtyard of the pagoda
11 Water purification (O)
12 Lodging (J3)
13 Lodging (H2)
14 Pool

0   6m

0   6m

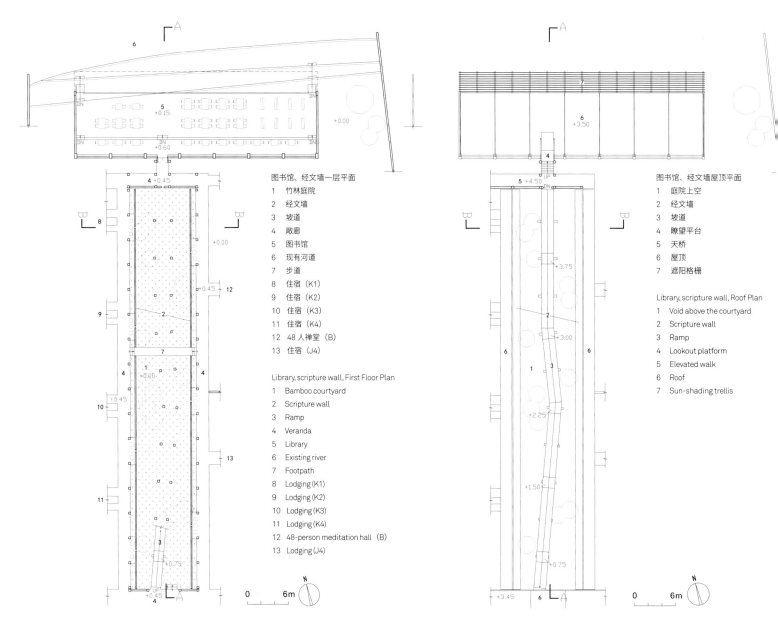

图书馆、经文墙一层平面

1　竹林庭院
2　经文墙
3　坡道
4　敞廊
5　图书馆
6　现有河道
7　步道
8　住宿（K1）
9　住宿（K2）
10　住宿（K3）
11　住宿（K4）
12　48人禅堂（B）
13　住宿（J4）

Library, scripture wall, First Floor Plan

1　Bamboo courtyard
2　Scripture wall
3　Ramp
4　Veranda
5　Library
6　Existing river
7　Footpath
8　Lodging (K1)
9　Lodging (K2)
10　Lodging (K3)
11　Lodging (K4)
12　48-person meditation hall（B）
13　Lodging (J4)

0　　　　6m

图书馆、经文墙屋顶平面

1　庭院上空
2　经文墙
3　坡道
4　瞭望平台
5　天桥
6　屋顶
7　遮阳格栅

Library, scripture wall, Roof Plan

1　Void above the courtyard
2　Scripture wall
3　Ramp
4　Lookout platform
5　Elevated walk
6　Roof
7　Sun-shading trellis

0　　　　6m

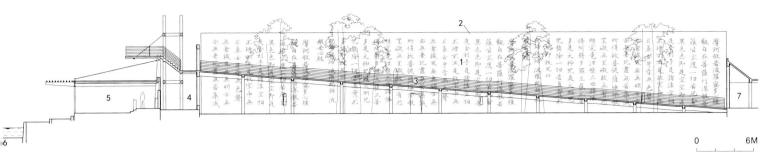

0   6M

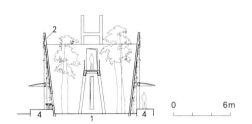

0   6m

（上）图书馆、经文墙 A-A 剖面，
（左）B-B 剖面

1 竹林庭院
2 经文墙
3 坡道
4 敞廊
5 图书馆
6 现有河道
7 飞地门厅（L）

(Above) Library, scripture wall, A-A
Section, (left) B-B Section

1 Bamboo courtyard
2 Scripture wall
3 Ramp
4 Veranda
5 Library
6 Existing river
7 Lobby for detached area (L)

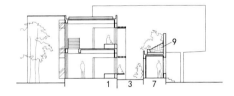

（下左）住宿（多人间 + 公卫）一层平面（局部），
（下右）二层平面（局部），（左）A-A 剖面

1 床榻
2 木平台
3 庭院
4 走廊
5 公共卫生间
6 门厅
7 敞廊
8 多功能室
9 屋顶绿化
10 遮阳格栅
11 庭院上空

(Lower left) Lodging (dormitory+public bath),
First Floor Plan (partial), (lower right) Second
Floor Plan (partial), (left) A-A Section

1 Bed
2 Wood deck
3 Courtyard
4 Corridor
5 Public bath
6 Lobby
7 Veranda
8 Multi-function
9 Planted roof
10 Sun-shading trellis
11 Void above courtyard

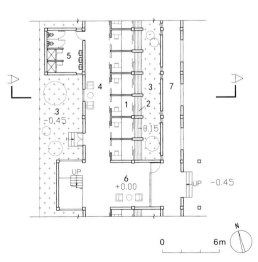

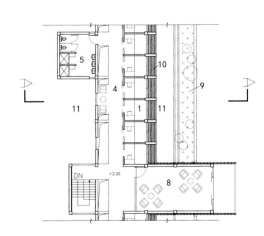

0   6m   N

湿地中的一艘"考察船"
——上海新江湾城生态展示馆（2005）

An "Exploration Vessel" in a Wetland
—New Jiangwan Ecological Exhibition Center, Shanghai (2005)

博物馆与生态保护区

说来人们可能不会相信，在今天上海中心城区边上，居然还存在着一片芦苇丛生、野鸟聚集的湿地。这片湿地是位于市区北缘的原江湾军用机场（始建于 1939 年）的一部分。在几十年的静置状态中，大自然逐步把这片土地演变成以若断若续的水面为主、辅以参天大树的世外桃源，与附近高楼林立的城区形成了令人难以置信的对比。在今天市区日益向外扩张的压力下，江湾机场的土地在 1997 年转为民用，被规划为占地 9.45 平方公里，含高等教育、居住、商务等多种功能的"新江湾城"开发区。非常幸运的是，这片湿地被保留了下来，成为新城中占地约 10 公顷的生态保护区，与开发区中的其他种植区域共同组成一个公共绿化系统。

为了避免干扰区内脆弱的生态系统，新城开发指挥部沿全区周边修建了隔声、隔光的围墙，并决定除了在向导带领下的小批量考察外，目前不向公众开放该保护区。但为了让公众同样能领略一点这片湿地的景色，了解生态保护的重要性，指挥部决定在湿地的西缘靠近新城主干道淞沪路的地方建设一座小型博物馆，除了展示有关该片湿地的生态知识以外，也为公众提供一个类似观察站的设施。

博物馆的建筑设计首先被纳入整个生态保护区的长期规划中来进行思考。我们认为，当经过一段较长的封闭期后，社会对生态保护重要性的理解更为普及时，应当可以容许公众通过架空步行道进入整个保护区。我们为此规划了两条高架步道，以博物馆为桥头堡，向保护区纵深方向辐射出去。未来的参观者可在保护区的东端走下步道，继续参观新江湾城公共绿化系统的其他部分，或沿另一条步道回到博物馆。在博物馆自身的设计上，我们在以下两个方向做了探索。

突破博物馆中传统的"黑匣子"空间

现代博物馆作为一种建筑类型，其室内通常采用一种与室外完全隔绝的"黑匣子"空间形式。这样一个展览空间必然完全依靠人工照明及气候控制，其展品也必然以标本等人工制品为主。这一空间概念当然有它的功能依据，如全人工控制的室内环境有利于保护如纸本绘画之类的展品，单一的环境形式有利于突出展品本身，并允许自由更换展览内容（图 1 左）。

但如果不盲目地接受现有权威，我们也可以说这种博物馆等同为仅仅是堆放展品的"仓库"的概念，毋宁说是反映了占西方思想方法主流的理性主义对明确分门别类的强调（如建筑对展品、室内对室外），特别是反映了早期现代建筑运动只看见生理功能、忽视心理功能的幼稚认识。

我们将发现"黑匣子"空间并不是没有它的功能缺陷的。首先，由于对建筑空间的体验必然是整个参观体验中的一个有机组成部分，将建筑背景与展览内容割裂开来就是放弃了一个有效的感染观众的手段。其次，将观众长时间地锁在黑匣子中，也不

符合人类喜欢亲近自然光与绿化的心理。事实上，无论是文艺复兴宫室中建筑对壁的烘托，或是中国传统园林中书画、匾额与景观的呼应，还是日本书院造民居中将挂古玩、插花及壁龛结合为一体的"床"，都揭示了另一种展览概念，它强调建筑空间与展品的互动，需要时还可为此将室内外结合在一起来形成展览空间。

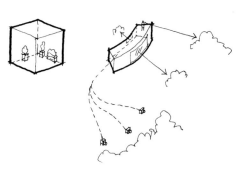

图 1. "黑匣子"展览空间（左）与室内外结合的展览空间（右）
Figure 1. "Black box" exhibition space (left) vs. indoor-outdoor exhibition space (right).

实际上，已经有少数国外建筑师在现代艺术博物馆中有限地突破了黑匣子一天下的局面，如耶尔·博（Jørgen Bo）与韦勒廉·沃赫莱尔特（Vilhelm Wohler）的丹麦路易斯安那（Louisiana）博物馆，查尔斯·柯里亚的新德里国家手工艺博物馆，或最近刚落成的赫佐格与德默隆的旧金山笛洋（de Young）博物馆。由于本案中主要展览的是附近的生态环境，展览布置又相对稳定，这使我们有可能在本工程中尝试一种与黑匣子完全对立的新的展览空间。

探索首先从最基本的设计概念出发。我们在建筑设计阶段就与业主共同确定了未来布展的大致形式与内容。由于这个工程的总建筑面积并不大，我们将整个博物馆的室内大半用作供观众移动的空间，主要展品均利用建筑壳体之外的生态景观形成。建筑设计的焦点定位在如何为参观者提供各种窗口来看成为展品的室外自然（图 1 右）。

为此我们首先在一层陈列厅设计了一个长近 13 米、高 1.8 米的带状半水下观察窗，透过窗能看见湿地水底、水体到岸边的一个完整剖面。我们按照本区的生态保护顾问德国 SBA 公司对当地三种湿地生态环境的界定设计了窗外的水体，使观众从窗口能看见涵盖浅水滩涂（与水面平）、中水区（水深 50 厘米）及深水区（水深 1 米）的典型生物、植被及地质构造的真实断面。（目前的室外水体是临时的，将在整个保护区全部水系后实现原设计。）

其次，我们在整个保护区内设置了五个摄像点。参观者可在一层的保护区模型上用激光笔随意选择感兴趣的区域，模型上方的大屏幕上就会实时显示出该区的

这种"看"的方式特别有利于用来观察不能受人干扰的生态环境，如鸟类活动等。

第三，在二层陈列厅的西墙外，我们设计了一个室外下沉庭院，人们可以通过长近 9 米、高 1.5 米的观察窗看到庭院中种植的园内典型植物。它们部分取代了...的标本陈列。

最后，屋顶层的瞭望平台为参观者直接观察区内陆上景观提供了一个 180°的全...面。为了丰富参观者的体验，我们在屋顶上设置了一座出挑 7.5 米的悬挑平台，...们可以就近观察水面对岸的树冠及鸟类。屋顶上局部升起的瞭望平台，不仅在...时方便了后排参观者，也为博物馆的室内自然通风提供了高窗的空间。

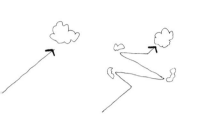

图 2. 瞬间揭示（左）与逐步揭示（右）。
Figure 2. Instant revelation (left) vs. gradual revelation (right).

...步揭示的参观路线

本设计的第二个探索方向是借鉴我国传统空间中常见的"逐步揭示"的概念。...建筑主流通常将建筑的室内与室外空间分成两个互相独立的系统，将所有室...间积聚成一个耸立在大片绿地中的雕塑体，建筑的表现力主要依靠瞬时间在...面前完全揭示这样一个雕塑的立面（图 2 左）。与此相反，我国建筑传统中...将建筑的室内外空间分成小块，将每个室内空间与一个室外空间配对，形成...各样的庭院或其他空间组合体。整个建筑的庐山真面目是在人们走过一个个...后被逐渐揭露出来的（图 2 右）。

保护区在独特的历史条件下形成，是现在上海市区内硕果仅存的"处女地"，...体报道后曾在市民中引起广泛关注。当相邻地区的建设完成后，保护区与周围...环境的对比将更为强烈，形成名副其实的"绿岛"。这些再加上全封闭式的管理，...相信保护区会很自然地在人们心目中形成"另一个世界"的意象。这种神秘感...化内风景对人们的感染力，提高人们对被保护环境的珍视度。作为市民得以...园内风景的唯一渠道，博物馆的设计应进一步加强这种神秘感而不是削弱它。"逐...示"正是达到这个目的一个有效措施。

但是，如何在一个总建筑面积不过约 370 平方米的工程中采用这样的手法呢？...的第一个策略是将整个建筑体处理成一个屏障。为此在基地靠近淞沪路入口处...了一个小型入口广场，条形的建筑体被放置在广场与园内水面之间，建筑被半...地下，向着淞沪路的一面做成台阶型，上覆土坡种树，与广场上的现有大树及...树连成一片。参观者从淞沪路望来将只看到一道小山坡，完全挡住了园内风景。...入博物馆，人们必须走下一列隐藏在遮荫花架下的台阶，才能进入半地下的一层。...喻了渔夫穿过黑暗的岩洞发现桃源胜境之意。

室内空间的基本组织原则是在建筑中部设置大空间的陈列厅，陈列的一二层之...有许多竖向空间联系。建筑南北两端为分隔成小空间的特殊或辅助功能，如视...、厕所及竖向交通。

一层陈列的主要吸引点是水下窗、保护区模型及从园内转播的摄像，所以从室...不见多少室外水面以上的实景，整个展厅沉浸在一片水下绿色的泛光中。

整个参观路线是"之"字形的。当人们从北向南看完一层展览后，在建筑南端...梯上到二层，从南向北继续参观。与较封闭的一层不同，二层朝园内的一面全...玻璃幕墙。但为了避免过早泄露园内景色的"天机"，整个幕墙外悬挂了一道

不锈钢丝网，日后将会爬满攀援植物。不仅如此，大部分幕墙玻璃内面还覆盖了一层带镂空图案的磨砂塑料薄膜，所刻图案同时又是展品之一———反映新江湾城历史上不同阶段的四幅地图。（按建筑师的原始设想，盛鸟类标本的玻璃盒应与幕墙脱开。）

当参观者跨过一座玻璃天桥，从二层北端楼梯来到屋顶平台上时，保护区的地面风景终于在人们眼前全部展开。随着人们在浏览中最后来到屋顶层南端时，整个参观活动逐渐进入尾声。参观者可经室外山坡上的台阶缓步回到入口广场。

在建筑师的心目中，这个博物馆就好像是停泊在湿地边的一艘科学考察船，人们在船中通过水下的、电子的、全景的等各种"眼睛"窥视着外面的大自然。常春藤将慢慢布满建筑东面玻璃幕墙外的金属丝网，屋顶上的花架也将逐渐被紫藤覆盖。当它们与建筑西面覆土上现在已经相当茂盛的乔灌木连成一片时，这艘"船"将消失在整个保护区的绿化之中。（2006）

Abstract

On the northern edge of Shanghai central city, a 10-hectare wetland had remained undisturbed for several decades until 1997 when it became a nature reserve and a part of the park system of a new town developed around the wetland. Since its fragile ecosystem cannot allow everyone to enter, it is necessary to establish an exhibition/observation center at its west side for the public. In the architect's long-term plan, two elevated public walkways will radiate from the exhibition center into the reserve.

Modern museum designs tend to follow a "black box" stereotype that isolates exhibition space from the outdoors. Considering the subject of the exhibition, this design instead makes the wetland surrounding the building the main exhibits. Resembling an exploration vessel, the building houses the visitors and opens up various windows for them to look at the live species and natural environment outside. On the first floor, a 13-meter-long underwater window shows a complete section of the wetland from its bottom to above the water surface. Adjacently, visitors can see real-time nature images, fed by five cameras in the reserve, by selecting a location from a model. A sunken courtyard on the second floor displays through a window live plants typical in the wetland. Finally, the observation deck on the roof affords visitors a panoramic view of the wetland.

The design borrows the idea of "gradual revelation" from traditional Chinese architecture to create a sense of mystery so that visitors will treasure the reserve even more. To people in the street the building appears as a small hill, blocking views of the reserve. After walking downwards into the half-buried first floor, visitors are attracted to the underwater window which makes people temporarily forget what happens above the water surface. On the second floor, the view of the reserve is screened by a layer of plant-covered mesh outside of the glass wall and the historic maps glued on the glass. Only after the visitors climb onto the roof deck, the scenery of the reserve fully reveals itself to them. People can then return to the entry plaza through a path on the slope of the man-made hill.

工程资料

地点 上海淞沪路（殷高路口）
时间 2004—2005
基地面积 2 420 平方米
建筑面积 362 平方米
业主 上海城投新江湾城工程建设指挥部

设计

建筑 / 展览设计概念：缪朴（上海市园林设计院顾问）
园林：庄伟（上海市园林设计院，下同）
结构：许蔓、陈彦、张业巍
给排水：茹雯美
电气：周乐燕
暖通：翼海风
展览设计：沈浩鹏设计工作室

发表 / 获奖

《建筑学报》（2006 年第 7 期），《时代建筑》（2006 年第 6 期）。
《建筑六十六》(2006)，《中国当代建筑 2004—2008》(2008)。
第一届上海市建筑学会建筑创作奖优秀奖，2006 年。
第七届远东建筑设计奖，入围，2010 年

Project Data
Location Songhu Road (at Yingao Road), Shanghai
Project Period 2004-2005
Site Area 2,420 square meters
Floor Area 362 square meters
Client Shanghai Chengtou (Division for New Jiangwan City)
Designer
Architecture/Concepts of Exhibition: Pu Miao (Design Architect), Shanghai Landscape Architecture De
Institute (Architect of Record)
Landscape: Zhuang Wei (Shanghai Landscape Architecture Design Institute, same below)
Structure: Xu Man, Chen Yan, Zhang Yewei
Plumbing: Ru Wenmei
Electrical Engineering: Zhou Leyan
HVAC: Yi Haifeng
Exhibition Design: Shen Haopeng Design Studio

Publication/Award
Architectural Journal (7/2006), *Time+Architecture* (6/2006).
Sixty Six-World New Architecture (2006), *Contemporary Architecture in China 2004-2008* (2008).
Award of Excellence, the First Architectural Design Awards, the Architectural Society of China (A
Shanghai Chapter, 2006.
Short-listed, the 7th Far Eastern Architectural Awards, 2010.

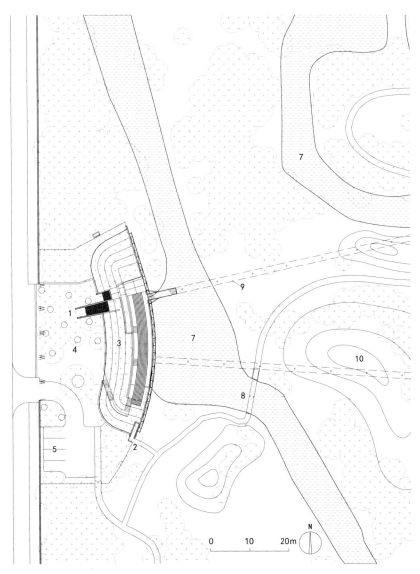

总平面
1 主入口
2 工作人员入口
3 山坡
4 小广场
5 停车
6 淞沪路
7 水面
8 吊桥
9 未来架空参观步道
10 生态保护区

Site Plan
1 Main entrance
2 Staff entrance
3 Hill
4 Plaza
5 Parking
6 Songhu Road
7 Water area
8 Drawbridge
9 Future elevated walkway
10 Nature reserve

（左到右）
二层及屋顶平面
主入口
商店
陈列厅
模型展坑
视听室
机房
储藏
杂务
工作人员入口
堆土
下沉庭院
厕所
瞭望平台
悬挑平台
小广场
水面
山坡

(From left to right)
First Floor, Second Floor and Roof Plans
1　Main entrance
2　Store
3　Exhibition
4　Pit for model
5　A/V room
6　Mechanical
7　Storage
8　Janitor
9　Staff entrance
10　Earth
11　Sunken courtyard
12　WC
13　Observation deck
14　Cantilevered deck
15　Plaza
16　Water area
17　Hill

殷高路　YINGAO ROAD

基地
SITE

生态保护区
NATURE RESERVE

淞沪路　SONGHU ROAD

0　　100m

N

基地位置平面
Site Location Map

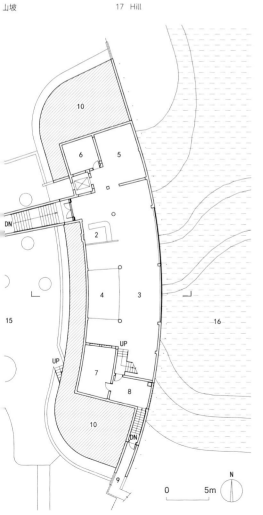

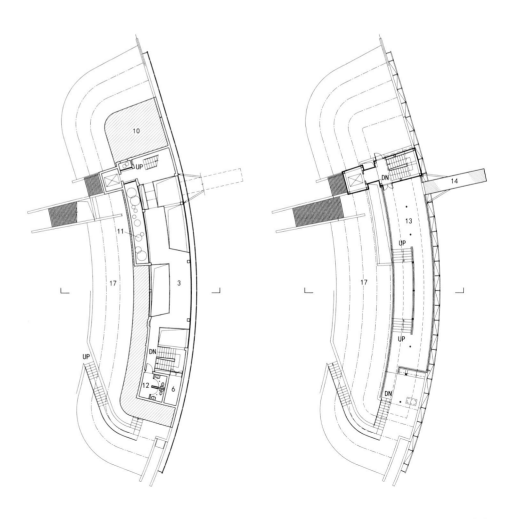

0　　5m　　N

建筑临街立面（西立面）像是一个长满树的小山
The street-side elevation (west elevation) look
a small hill covered by trees.

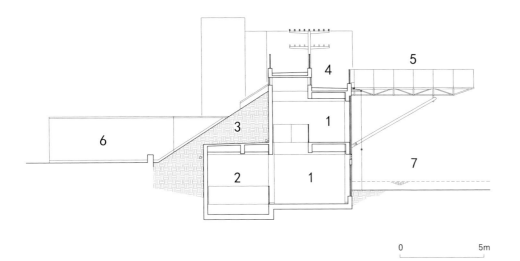

剖面
1 陈列厅
2 模型展坑
3 堆土
4 瞭望平台
5 悬挑平台
6 小广场
7 水面

Section
1 Exhibition
2 Pit for model
3 Earth
4 Observation deck
5 Cantilevered deck
6 Plaza
7 Water area

0   5m

幽暗的下行入口暗示进入"另一个世界"。
Shaded and downward entrance suggests entering "another world."

从入口望一层陈列厅中的水下观察窗。
From the entrance looking toward the underwater windows in the first-floor exhibition.

一、二层陈列厅的空间垂直贯通
Vertical penetration between exhibition spaces on the first the second floors.

层陈列厅南端北望。

the south end of the second-floor exhibition

ng north.

二层陈列厅东面玻璃幕墙上的历史地图及鸟类标本。

The maps and bird specimens on the glass curtain wall along the east side of the second-floor exhibition.

一层陈列厅西边的保护区模型及与之互动的园内实况转播屏幕。

On the west side of the first-floor exhibition are a model of the nature reserve and monitors (interactive with the model) showing videos transmitted from the reserve.

二层陈列厅西面的植物观察窗。
Windows to observe live plant exhibits on the
west side of the second-floor exhibition.

顶层悬挑平台回望楼梯间。

From the cantilevered deck on the roof looking toward stair tower.

建筑临湿地立面（东立面）中段。
The middle part of the wetland-side elevation (east elevation).

建筑东立面上的悬挑平台。
The cantilevered deck in the east elevation.

建筑西南角的下山台阶。
The downhill steps at the southwest corner of the building.

"楼梯"建筑
——洛阳小浪底公园茶室（2002）

"Stair" Building
–Teahouse, Xiaolangdi Dam Park, Luoyang (2002)

黄河经过河南洛阳附近的小浪底大坝后，从原河道北面与之平行的新辟水道泄出。坝脚下的一段原河道除了为水坝排泄正常渗水外，其周围 52.6 公顷的区域被开发为公园。按照上海市园林设计院周在春先生的规划，部分河道被拓宽为湖面，并在湖与河道相交处安排了一座茶室。从该处西望，正对大坝及其脚下的湖中岛屿，向北是供游泳者使用的湖滨沙滩，东面为高十多米的小山丘，山后为顺一道溪流建造的黄河水利工程模型，沿基地南面全长则是原来的黄河。我有幸受周先生邀请与园林院同仁提供了茶室的设计。在甲方代表陈玲女士的认真监督下，本工程在 2002 年秋基本按计划建成。

该茶室由四个条形体块组成。其中两个在立面上呈曲线升起的体块构成茶室的主体，内含 11 个楼面标高从 1.00 米依次递增到 11.00 米的餐厅 / 茶座。每个餐厅可坐 80 人，全部面对大坝及湖景，但高度的变化使各个餐厅的景观从以湖中曲桥、小山过渡到以雄伟的大坝为中心。除了北端的三个外，所有餐厅均仅用矮墙分隔，从室内形成一个连续的台阶型空间。在此"大台阶"的标高 5.00 米及 10.00 米处，有一道天桥延伸到建筑背后的小山山腰及山顶上，从下面上来的游人可由此继续游览山后的其他景观（从山后过来的游人则可探索湖景）。

第三个条形体块平卧在建筑北面，内含供游泳者使用的更衣室、室外淋浴小院等。第四个体块从上述两个曲线体块之间垂直升起，内设分别供游人及服务员专用的楼梯、食梯及备餐间等。食梯及备餐间的位置安排使服务员最多只需爬相当于普通建筑中的一层楼就可以到达最远的餐桌。楼梯间的脚下为厨房，其南设有一处供残疾人使用的餐厅。

两个曲线体块的建筑形式为露明的钢筋混凝土梁柱，朝大坝一面的填充墙全部为清或磨砂玻璃。柱身设在玻璃面的外面，期望能对西晒起一定遮阳作用。另两个体块则采用贴面砖的实墙，辅以铝窗。

由于此公园在夏季的游人远远超过其他季节，本设计利用上述四个体块的互相穿插在建筑的西、北、南三面界定出三个高度不同的室外用餐区域：建筑西面的室外座可容纳大量的快餐顾客；更衣室体块屋顶上的咖啡座可供游泳者穿着泳衣随意小坐；建筑南面的一个小型屋顶餐厅不仅提供了安静的气氛，同时正对原来的黄河。

上述的设计揭示了本建筑既是连接小山上下两个景区的一个大楼梯，又是一个供人休息的茶室。游人走过 11 个依次升高的茶座平台时，可以随处坐下来欣赏从坝脚小山上升到大坝全景的各个画面。在大楼梯的中间和顶端的两个天桥，将游人最终带到背后的小山上。就像颐和园里的长廊那样，这座园林建筑让游览路线穿过它自己，让人能从"景中"到"景外"（指从室内往外望时），再回到"景中"来看风景，从较长的一段路线中得到更丰富的体验（冯纪忠先生最早提出这个理论）[1]。这与园林传统里的建筑大多集中一处，并主要是被当作雕塑来欣赏其外观相比，有着本质上的不同。

"楼梯"建筑其他人也做过，像环形的台阶式空间最早被应用在赖特的纽约古根海姆美术馆及芦原义信的东京索尼大楼。近年来则有雷姆·库哈斯（Rem Koolhaas）在荷兰乌特勒支（Utrecht）大学讲堂中试验了平行的楼梯空间。但这些都是平地上的房子，走上去了还得再走下来。这次与小山结合在一起做，游人将会感到更自然吧！

台阶式的室内空间同时在建筑外形上提供了尝试一种另类构图体系的机会。在现在流行的构图语言中，无论是坚守规则几何形，还是偏爱自由曲线，构图中各个形体之间的关系总是强调对立甚至碰撞。但为什么不试试从直线中逐步演化出的"平缓曲线"呢 [2]？当然，后者让人感到的是一种含蓄的变化而不是震惊，但为什么只有让人吓一跳才是美的呢？

建筑的形状出来了还得考虑颜色。原来曾打算套用国际上现在流行的本色混凝土梁柱加白墙，但一看周围的黄土峭壁及青翠植被，还有在当地民居中大量使用的红砖墙，这样的色彩恐怕会显得苍白。最后为框架选择了一种特别的红色，填充墙（东立面上）则用了带紫的深蓝（后者在施工时做得淡了一点）。这些在洛阳白马寺中看到的颜色既鲜艳又不同于商业广告中那类很纯的原色。夕阳西下时，从建筑南面的屋顶餐厅望黄河原河道，在茶室暗红色的梁柱衬托下，河对岸古老的黄土峭壁及苍翠的草木显得分外夺目。写到这里不由想到在克里斯朵夫·亚历山大（Christopher Alexander）的色彩课上听他说的那种"强烈但安静"的感觉，那都是十多年前的事了。（2004）

注释

[1] 冯纪忠，《组景刍议》，《同济大学学报》1979 年第 4 期（建筑版），第 1-5 页。

[2] 缪朴，《传统的本质——中国传统建筑的十三个特点》，《建筑师》第 40 期（1991 年 3 月）第 68 页及《台湾大学建筑与城乡研究学报》第 5 卷，第 1 期（1990 年 2 月），第 68 页。

Abstract

The main purpose of the Xiaolangdi Dam Park is to allow tourists to view the largest dam on the Yellow River. People can also swim in a lake made out of the original river course. The teahouse is located between the lake (west of which looms the dam) and a hill. Behind the hill lie other attractions of the park.

The teahouse has four components. The two curved ones contain 11 tea rooms, each sits at a different level. The third component lies at the northern side to serve the swimmers. The fourth component, a tower, houses stairs and a dumbwaiter. At the bottom of the tower are the kitchen and a dining room for handicapped visitors. Several outdoor dining areas are created for the additional summer users.

In contrast to the European gardens which treat buildings as isolated sculpture, the design integrates architectural space into the landscape touring experience, a concept borrowed from the Chinese garden tradition. The tea-room wings are actually two continuous "stepped spaces." Two bridges connect the stepped spaces to the middle and the top of the hill behind the building. Tourists from either the hill or the lake can continue their trip through the teahouse or pick a tea room to sit down. The changing heights of tea rooms afford people different views that move from the dam to the lake. Thus the building becomes part of the tour path of the park.

In the building form, the design explores "flat curves," a unique geometry observed in Chinese calligraphy and traditional architecture. The gentle curves grow out of straight lines, suggesting both a variation from and continuity of the orthogonal form. The geometry differs from the Western formal tradition which emphasizes self-complete curves and a contrast between curved and orthogonal geometries. The colors of the building are based on the popular tones used in local farmers' houses and temples. They are bright but never pure, appearing both strong and silent. Such colors work well with the surrounding hills of yellow earth and dark green trees.

工程资料

地点 河南省洛阳市小浪底大坝公园
时间 2000—2002
建筑面积 5 382 平方米
业主 小浪底建设管理局
设计
建筑：缪朴（上海市园林设计院顾问）
结构：徐瑞倩（上海市园林设计院，下同）
给排水：陈惠君
电气：周乐燕
公园规划：周在春

发表
《建筑学报》（2003 年第 3 期），《时代建筑》（2004 年第 4 期）。

Project Data
Location Xiaolangdi Dam Park, Luoyang, Henan Province, China
Project Period 2000-2002
Floor Area 5,382 square meters
Client Xiaolangdi Bureau of Construction Administration
Designer
Architecture: Pu Miao (Design Architect), Shanghai Landscape Architecture Design Institute (Architect of Rec
Structure: Xu Ruiqian (Shanghai Landscape Architecture Design Institute, same below)
Plumbing: Chen Huijun
Electrical Engineering: Zhou Leyan
Park Planning: Zhou Zaichun

Publication
Architectural Journal (3/2003), *Time+Architecture* (4/2004).

一层平面
1 餐厅 / 茶座
2 自助餐柜台
3 室外餐厅
4 厨房
5 电气
6 广播
7 卸货平台
8 男 / 女更衣
9 厕所
10 室外淋浴

First Floor Plan
1 Dining/tea room
2 Buffet counter
3 Outdoor dining
4 Kitchen
5 Electrical
6 Broadcasting
7 Loading deck
8 Men's/women's locker room
9 WC
10 Outdoor showers

0   5   10m   N

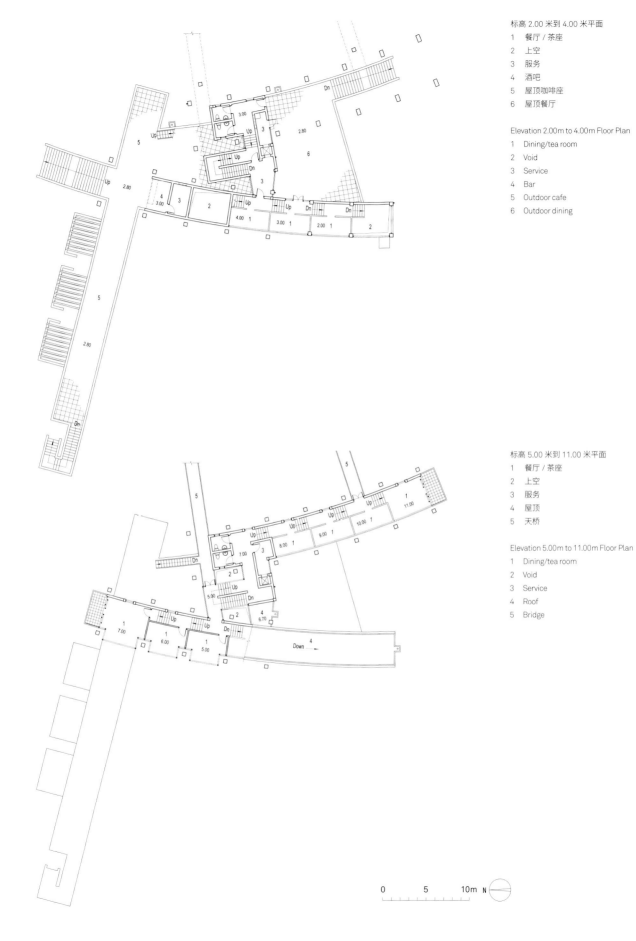

标高 2.00 米到 4.00 米平面

1 餐厅 / 茶座
2 上空
3 服务
4 酒吧
5 屋顶咖啡座
6 屋顶餐厅

Elevation 2.00m to 4.00m Floor Plan

1 Dining/tea room
2 Void
3 Service
4 Bar
5 Outdoor cafe
6 Outdoor dining

标高 5.00 米到 11.00 米平面

1 餐厅 / 茶座
2 上空
3 服务
4 屋顶
5 天桥

Elevation 5.00m to 11.00m Floor Plan

1 Dining/tea room
2 Void
3 Service
4 Roof
5 Bridge

0   5   10m   N

西面全景。
West view of the entire bui●

小岛
ISLAND

湖
LAKE

沙滩
BEACH

134

150

137

机房
MECH.

141

137

茶室
TEAHOUSE

±0.00=139

亭
PAVILION

150

小溪
STREAM

桥
BRIDGE

黄河故道
OLD COURSE OF THE YELLOW RIVER

137

0   30   60m

N

总平面
Site Plan

室翼屋顶咖啡座。

oor cafe on the roof of the
er-room wing.

画局部外景，更衣室翼大
画面外，楼后小山绿化尚
。

l southwest view, locker-
wing mostly invisible,
g of the hill not completed

南面局部外景。
Partial south view.

南面局部夜景。
Night view of the south side.

由南面屋顶餐厅透过台阶空间上门洞望湖中小岛
View from the southern roof terrace toward the
looking through an opening in the stepped volur

由东北角天桥西望更衣室翼屋顶咖啡座。

From the bridge at the northeast corner looking west toward the roof terrace of the locker-room wing.

空间中的一个餐厅／茶座。

of the stepped dining/tea rooms.

面屋顶餐厅望黄河对岸峭壁。

n the southern roof terrace looking toward the
w River and the cliff behind.

与水对话
——昆山思常路茶室（2007）

Conversing with the Water
—Sichang-Road Teahouse, Kunshan (2007)

当前我国城市建设中出现的"千人一面"现象，已经引起社会上广泛的注意。其中一个原因，是城市与建筑的形式与当地环境的脱节。比如说，在大规模工业化进程开始前，纵横交错的港汊芦荡是江南水乡城镇的典型风景。传统的城镇布局以及建筑设计在几百年中与河道形成了独特的亲密关系，建筑可以沿河而造，或者悬挑在河边，甚至整个架空在水面上。但在最近30年的城市更新及扩展中，大量河道被填埋，与之一同消失的是那种与水共生的河边建筑。开发商虽然在填河产生的土地上又挖了一些水池，但不少是与基地外自然水系没有关系的一潭死水。最成问题的是，绝大多数所谓的"亲水"建筑，离水总是至少有三尺远。这既可归咎于死板的政府法规（特别是在城市规划的水体边上时），但更根本的原因则来自设计者不顾地形、照搬成思路（如所谓"欧陆风格"之类）的做法。我们在不少实例中可以发现，建筑师实际上不过是将一个在旱地上的普通别墅设计放在附近有河的位置上罢了，建筑根本没有与水发生真正的互动。结果，在水泥森林及人造"景观"中长大的城市居民对自己的乡土风景记忆越来越淡漠。就好像对蚯蚓、螳螂很陌生一样，城市儿童也很少有机会接触到自然状态的水。这样的环境是不是城市"现代化"的必然结果呢？针对这个课题，笔者一直在找机会探索另外的设计途径。

江苏省昆山市是一个大家都知道的江南水乡古城，但经过1980年代以来的多次改建更新，这个城市的物质面貌已经变得与周围成百上千个中小型城镇差不多了。上面说的问题在这里也都存在。在最近三年里，一个占地80余公顷、人口3万的小区——康居小区在昆山西郊逐渐成形。非常值得庆幸的是，康居小区的规划中保留下一些天然河道，本文介绍的项目就坐落在河边。康居小区将主要用来安置因改造老城而产生的拆迁居民，所以特别强调要创造完备成熟的居住环境。小区内规划了系统的商业设施及公共绿地，思常路茶室就是该公共配套设施的组成部分之一。因此，如何在河边茶室中重新体现人与河之间的亲密关系，增加城市居民与自然环境的交往，成了本设计的题中应有之义。

思常路位于该小区的东北角，与一条南北走向、约25米宽的河道平行。河与道路之间有一条宽四五十米、长800余米的线形公共绿地。园林设计师在该绿地北端（靠近马鞍山路桥处）规划了一个茶室。基地面临的思常路另一边将建设大型商业设施，所以，位于沿河绿地中的这个小建筑，将为在百货大楼或超市中逛累了的居民们提供一个另类的接近自然的休闲场所。

为了最大程度地利用沿河景观面，建筑基本取线形体态。从道路到河边由四个平行的空间层次组成：（1）沿人行道是一道上植树木的小山坡，将建筑隐藏其后，从道路上望来避免了在视觉上打断沿河公共绿带；（2）躲在小山坡后的是一道种满竹子的庭院，为建筑西面提供了采光、通风及景观的可能；（3）沿竹院全长是一条室内走道，其南端开向一个上有花架的小型室外空间，需要时也可用来招待更多顾客；（4）像一串念珠那样，走道在临水一侧连接了各个茶座空间。与上述建筑主体的水平条状开对比，茶室的主入口被处理成唯一的垂直体块，突出在主体之外。以下介绍一下本计的两个目标：使用者与水之间的互动，以及对"平缓曲面"的采用。

人—河关系

要实现上述亲水目的，第一个阻碍是天然河道的水位上下落差很大。如何使们能总是接近会变动的水面呢？我们为此在建筑与河道之间设计了一个窄长的中介池，池水取自河中。水池邻河的一边边缘做了特殊处理，所以从茶室内望去，池水河水融为一体。我们在水池中断断续续种植了一列水杉，这样不仅从河对岸望来延了现有沿河的水杉林带，而且在从茶室内望河的画面中增加了一个空间层次。

拉近人—河关系的另一个手法是在平面上做文章。本工程中的茶座被分成散座包厢两大类。每个包厢（四人一桌）均设置在一个被水环绕的玻璃圆柱体中，圆柱内的地面是半沉在水中的。圆柱体上有大窗可以打开让人触及水面。由于水面就的胳膊肘下，这让每个坐在下沉茶座中的茶客均有身处舟中之感。各个下沉茶座的水面中安置了多个涌泉，给这些狭窄的水面带来生气。视线穿过圆柱体的玻璃屋人们可以看见上面还有一层木花架。当花架上爬满藤蔓类植物时，茶客们会享受处树荫中，从被庇护的暗处眺望河面明亮风景的那种舒适感。与包厢四周是水的不同，位于建筑北端的散座厅采取传统的一面临水的形式，但地面与水面接近，了反光的黑色自流平涂料，使人感到是水面的延伸。

平缓曲线

除了人与河的关系外，本设计的第二个目标是在建筑实体的几何形式上探索新的语言。我国传统建筑形式中通常含有一种"平缓曲线"，它们是正交直线体系延伸微变而不是对立，从而在建筑总体效果上形成一种含蓄的变化，而不是国外时髦风格的那种碰撞叫嚣之感。为了使建筑与小河在几何形式上更好地联系起来，们在河边茶室的花架屋顶中实验了上述概念。出于强调本建筑现代感的目的，这面被设计成三维的。为此我们特地探索了如何用当地施工单位能理解的二维图纸及施工技术，来实现这一三维曲面。具体做法是将该面理解成由两条不平行导线成的曲面，支撑屋面的八榀钢筋混凝土框架则成为曲率的控制点。因此，各榀框部分尺寸都不一样。但与程控切割的钢结构相比，这不需要复杂的技术与昂贵的设制作混凝土模板的人工成本相对上述高技做法来说还是较低的。

记得过去在哪里读到一个19世纪的西方作者回答这样一个问题：为什么欧市个个不同，而美国城市却看上去都一个样？他说这恐怕是因为美国的城市相对

很年幼，就好像婴儿们的脸那样个个相似。今天中国的城市恐怕都不用担心自己缺
史或文化吧，但在城市面貌上却为什么都出现了病态的返老还童症呢？包括思常
室在内的昆山的这几个小小实验，正是试图在建筑层面上建设性地回答这个问题。
009)

tract

he Yangtze River Delta region where Kunshan sits, towns historically developed
ique landscape of crisscross canals and numerous ponds. Traditional buildings
next to or even cantilevered over the canals. Unfortunately, the urban
ewals in the past three decades have transformed many urban natural water
as into streets and building sites. Regulations further keep people and buildings
any close contact with remaining rivers and lakes. As a result, today's urban
dren often forget the feeling of natural water, along with earthworms and
kets, which are part of the symptoms of apathy toward nature in Chinese high-
sity cities.

ated between a street and a preserved natural river in a new residential area,
design of the teahouse tries to address the above problem by encouraging
ple to develop intimacy with natural water. The river level fluctuates greatly.
refore the design creates an intermediating pool that draws its water from
river. Viewed from the teahouse, the pool appears to merge with the river. A
of *Metasequoia* trees along the river bank continues into the pool, providing
ground to the river view. Ten private tea rooms take the form of glass pods
ounded by water and half-sunken into the pool. Users can open the windows
touch the water under their elbows, just like in a boat. Tiny fountains (drawing
er from the river) are bubbling in the gaps between the pods. A layer of wood
lises with vines above the glass roof affords people the comfort of peeking into
bright river from under dark "tree" shades.

design also experiments with the "flat curve" observed in traditional Chinese
hitecture. As a gradual transformation of the orthogonal composition rather
n a sharp contrast to the latter, the flat curve creates a subtle variation instead
oisy drama. Constructed with a low-tech and low-cost method, the trellis roof
pts such a form in double-curvature to better relate the building with the river.

工程资料
地点 江苏省昆山市思常路马鞍山路口
时间 2005—2007
建筑面积 约 300 平方米
业主 江苏省昆山城市建设投资发展有限公司
设计
建筑及室内装修：缪朴（上海市园林设计院顾问）
园林：庄伟（上海市园林设计院，下同）
结构：许曼
给排水：李雯
电气：周乐燕

发表
《建筑学报》（2009 年第 1 期），《新建筑》（2012 年第 3 期），archdaily.com。
《建筑'09'08——当代中国建筑创作论坛作品集》(2009)，《当代中国建筑地图》(2014)。
"水——诅咒还是祝福！？亚太地区前瞻性建筑"展览及讨论会，德国柏林 Aedes East 当代国际建筑论坛，
2011 年。

Project Data
Location Sichang Road at Maanshan Road, Kunshan, Jiangsu Province, China
Project Period 2005-2007
Floor Area 300 square meters
Client Kunshan City Construction, Investment and Development Co., Ltd.
Designer
Architecture: Pu Miao (Design Architect), Shanghai Landscape Architecture Design Institute (Architect of Record)
Landscape: Zhuang Wei (Shanghai Landscape Architecture Design Institute, same below)
Structure: Xu Man
Plumbing: Li Wen
Electrical Engineering: Zhou Leyan

Publication
Architectural Journal (1/2009), *New Architecture* (3/2012), archdaily.com.
Architecture'09'08: Works from the Forum on Chinese Contemporary Architecture (2009), *Atlas of Contemporary Chinese Architecture* (2014).
Exhibited at the "Water—Curse or Blessing!? Encouraging Architectural Projects in Asia-Pacific," an exhibition and symposium at the Aedes East-International Forum for Contemporary Architecture, Berlin, Germany, 2011.

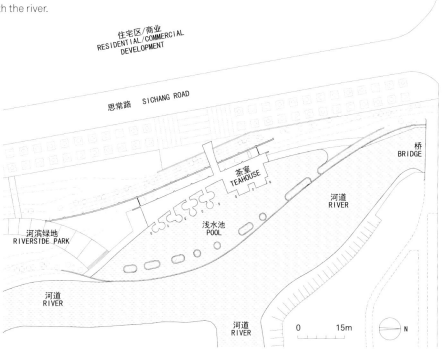

总平面
Site Plan

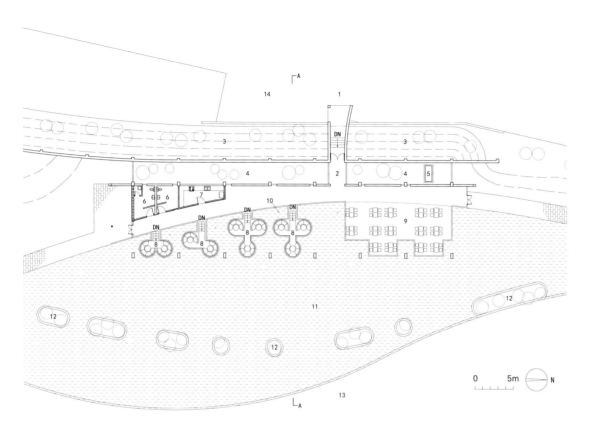

一层平面
1 主入口
2 门厅
3 堆土
4 庭院
5 空调设备
6 男 / 女厕
7 厨房
8 下沉茶座
9 散座茶座
10 涌泉
11 浅水池
12 树池
13 河道
14 人行道

First Floor Plan
1 Main entrance
2 Foyer
3 Mound
4 Courtyard
5 AC equipment
6 WC
7 Kitchen
8 Sunken tearoom
9 Large tearoom
10 Fountain
11 Pool
12 Tree pit
13 River
14 Sidewalk

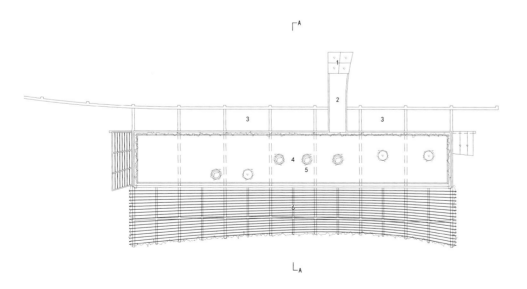

屋顶平面
1 入口标志
2 屋顶
3 庭院上空
4 植草屋顶
5 天窗
6 花架

Roof Plan
1 Entrance sign
2 Roof
3 Void above courtyard
4 Planted roof
5 Skylight
6 Trellises

东北角，中介水池及河。
east corner of the building, the intermediating
nd the river.

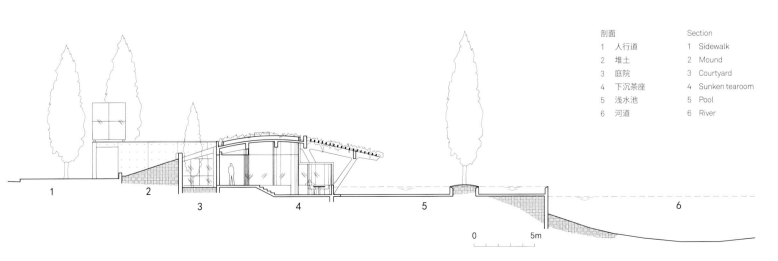

剖面		Section	
1	人行道	1	Sidewalk
2	堆土	2	Mound
3	庭院	3	Courtyard
4	下沉茶座	4	Sunken tearoom
5	浅水池	5	Pool
6	河道	6	River

0   5m

从下沉茶座望散座茶座。
Looking toward the large tearoom from the sunken tearooms.

东南角外观。
heast view of the building.

下沉茶座。
Sunken tearooms.

从走廊内望下沉茶座之间的涌泉。
From the corridor looking toward a founta
between the sunken tearooms.

作为公共空间的餐厅
——昆山思常公园餐厅（2012）
Restaurant as a Public Space
—Restaurant, Sichang Park, Kunshan (2012)

餐厅、茶馆、俱乐部等这类建筑有一种独特的两重性。它们既是商业设施，又是公众聚会的场所；既是一个私有的领域，又是城市公共空间系统的一部分。由于亚洲城市人口密度普遍较高，大多数家庭居住在公寓里，不但室内空间有限，也不像在北美普及的独立住宅那样有宽敞的前后院，可以让室内活动延伸出去。所以，对亚洲城市的中下层收入的居民来说，接待访客、大家庭团聚或几个邻居在一起打牌看球赛，都不大可能像西方居民那样在宽敞的客厅、家庭室、车库或后院中进行。无论是在东京还是首尔，我们发现这些活动通常被转移到城市公园、邻里儿童游戏场或像咖啡馆这样的半商业半公共设施中。

我国目前正当社会转型，许多原来由政府提供的纯公共福利设施（如工人文化宫等）逐渐失去财政支持，而像成熟工业化社会中由民选政府保证的公共服务尚未制度化。因此，以商业为主兼具社交功能的场所将在我国目前的城市生活中起到越来越重要的作用。如何在公私两种功能之间取得平衡，公园餐厅实在是一个可以探讨的有趣建筑类型。

昆山是上海西面一个蓬勃发展中的工业城市。本建筑位于该城新开发的郊区，坐落在一个在保留下来的自然河道两侧形成的线形公园内，公园的东西两面分别面临一所学校及一个待建设的公寓工程。

建筑的基本形状为南北长、东西窄的长条形，沿南北走向的河道东岸平行布置，保证了尽可能多的就餐者可以眺望河景。长条形的东西两面处理截然不同。东面基本为一道实墙，面水的西面主要为透明玻璃，强调了建筑向水面开放的姿态。

公园的景观设计师为公园设计了一个城市农业的主题。在建筑基地东面的区域中布满了东西走向的条形菜圃。建筑设计让这些菜圃好像"穿入"建筑，在室内变成玻璃围合的条形庭院。庭院上设A字形钢管花架，花架与庭院地面之间挂绷紧的不锈钢缆，上爬耐阴攀藤植物。条形庭院在建筑西面又延伸到室外，成为条形种植坛。靠近建筑的种植坛设与条形庭院相似的花架及钢缆，可种植丝瓜、西红柿等攀藤蔬果，其余段落的种植坛可种植不需攀藤的蔬菜。种植蔬菜既满足了公园规划中原来设计的科普教育功能，同时让顾客有"看什么，吃什么"的新鲜就餐体验。

但更重要的是，条形庭院及种植坛将一层及屋顶平台上的就餐区域分隔成多个半私密空间。花架上的攀援植物就像疏松的绿色屏风，能带来一定的私密感但不会让半私密的空间彼此完全孤立。餐厅给食客的独特体验之一是能与其他人或其他小团体共享一个有意思的就餐空间。目前中国常见的餐厅设计倾向于将大部分就餐面积设计成全隔断的包间，使餐厅大大削弱了它作为公共空间的交往功能。这一趋势当然反映了今天中国社会对公共生活更深层的看法，但本设计尝试用"绿帘"代替实墙，在交往与私密需求之间争取平衡。这是我们用建筑手法振兴公共空间的一个尝试。

中国传统建筑的基本概念之一是将每一个室内空间与一或多个室外空间"配对"，来服务于一个建筑功能，这是本设计所试图继承的。一、二层的室内就餐空间均与室外平台或屋顶平台相邻，人们可以随意在室内或室外环境中选择自己喜爱的地方进○邻近的爬藤花架使顾客能始终在绿化环抱中就餐，提供了一个既有室内的舒适、又○自然景观的丰富环境，有别于西方建筑中大多分离室内外空间的做法。重重东西走○的爬藤花架，也为面河的建筑西立面在夏天清晨及傍晚提供了必要的垂直遮阳。(20

Abstract

The restaurant is located in a linear park along a preserved natural river in a ne developed suburban area of Kunshan. A school and a future apartment proj flank the east and west sides of the Sichang Park.

The river-front building has distinctively different designs for its west (facing river) and east facades. A mostly solid wall on the east indicates that the buil opens up toward its river side, where transparent surfaces dominate.

The landscape architect adopted an urban agriculture theme for the park. Str of vegetable beds were laid out in the areas along the east side of the building The architectural design made the strips appear to "penetrate" into the build Once inside, the strips become glazed linear courtyards, dividing the first-fl dining area into semi-private spaces. The climbing plants in the linear courtya create the effect of green latticed screens that provide certain privacy but do completely separate the semi-private spaces from each other. The design a to avoid the over-compartmentalized dining rooms popular in today's Chin restaurants, promoting the restaurant as a public space.

A two-story-high foyer welcomes diners to the second floor. On both floors, th are outdoor terraces or roof decks next to the indoor dining spaces. People easily choose their favorite dining spots between indoor and outdoor settings. adjacent plant-climbing A-frames provide not only the visual pleasure of nat but also the sun shading needed in the summer. Here the design revives a tradi in Chinese architecture that provides a combination of indoor and outdoor setti for each activity.

资料

: 江苏省昆山市思常路（前进西路口）

: 2009—2012

面积 569 平方米

: 江苏省昆山城市建设投资发展有限公司

: 缪朴设计工作室（缪朴）；汉嘉设计集团（蒋宁清）

: 上海源规建筑结构设计事务所（张业巍、梁栋）

: 汉嘉设计集团（郭忠、于洋、吴秋燕）

chitectural Review》（2012 年第 9 期），《时代建筑》（2012 年第 6 期）。

Project Data
Location Sichang Road at West Qianjin Road, Kunshan, Jiangsu Province, China
Project Period 2009-2012
Floor Area 569 square meters
Client Kunshan City Construction, Investment and Development Co., Ltd.
Designer
Architecture: Miao Design Studio (Design Architect), Pu Miao; Hanjia Design Group, Shanghai (Architect of Record), Jiang Ningqing
Structure: Shanghai Yuangui Structural Design Inc., Zhang Yewei, Liang Dong
Engineering: Hanjia Design Group, Shanghai, Guo Zhong, Yu Yang, Wu Qiuyan

Publication
Architectural Review (9/2012), *Time+Architecture* (6/2012).

总平面
1　餐厅
2　菜圃
3　河道
4　思常公园
5　辅助建筑
6　思常路
7　前进西路
8　学校
9　住宅区

Site Plan
1　Restaurant
2　Vegetable beds
3　River
4　Sichang Park
5　Auxiliary building
6　Sichang Road
7　West Qianjin Road
8　School
9　Housing

0　50　100m

N

一层平面
1　门厅
2　半私密就餐空间
3　条形庭院
4　室外就餐平台
5　厨房
6　食梯
7　男 / 女厕
8　河道
9　芦苇
10　条形种植坛
11　菜圃

First Floor Plan
1　Foyer
2　Semi-private dining space
3　Linear courtyard
4　Outdoor dining terrace
5　Kitchen
6　Dumbwaiter
7　WC
8　River
9　Reeds
10　Linear planter
11　Vegetable beds

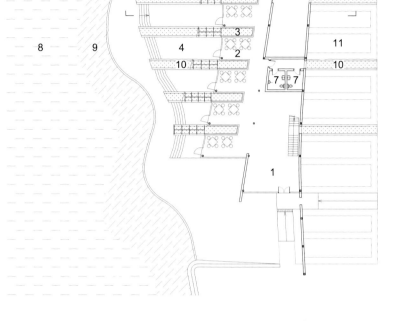

二层平面
1　大餐厅
2　屋顶平台
3　上空

Second Floor Plan
1　Large dining space
2　Roof deck
3　Void

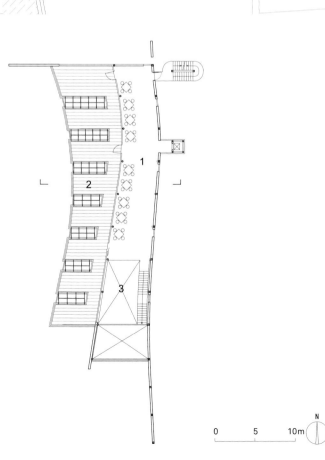

0　　5　　10m

N

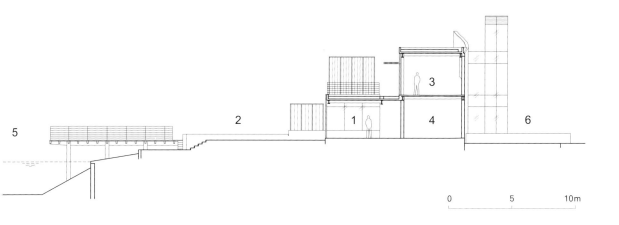

剖面
1 半私密就餐空间
2 室外就餐平台
3 大餐厅
4 厨房
5 河道
6 条形种植坛

Section
1 Semi-private dining space
2 Outdoor dining terrace
3 Large dining space
4 Kitchen
5 River
6 Linear planter

0   5   10m

西北面外观。
hwest view of the building.

从西南角望建筑西面由 A 字形爬藤花架分隔的室外
平台。

From the southwest corner looking toward th
decks separated by plant-climbing A-frames alon
the west side of the building.

建筑西南面外观。
Southwest view of the building.

建筑东北外观。
Northeast view of the building.

南端的主入口。

main entrance at the south end of the building.

（左）种植坛好像穿入建筑东面。

(low left) A planter seems to penetrate the elevation of the building.

（右）门厅。

(low right) Foyer.

从门厅望由条形庭院分隔的小餐厅。
From the foyer looking into the dining spaces
separated by linear courtyards.

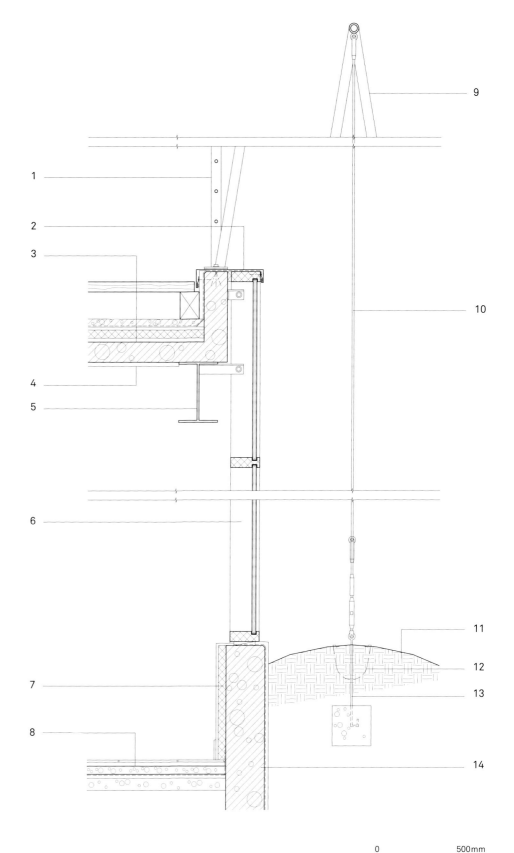

条形庭院外墙剖面

1　镀锌钢栏杆，做涂料
2　2.5mm 铝板压顶
3　屋面构造：
　　防腐木甲板；40mm 配筋细石混凝土；油毡；
　　屋面防水层；15mm 水泥砂浆；40mm 硬质
　　保温层；轻质混凝土找坡；现浇钢筋混凝
　　土屋面板
4　水泥石灰砂浆天花，白色涂料
5　钢梁外做防火层及装饰涂料
6　铝框幕墙，双层保温玻璃及 low-E 镀层
7　XPS 保温板带石膏板面层，做涂料
8　地面构造：
　　陶土地砖；20mm 水泥砂浆；40mm 细石
　　混凝土；防水涂料层；20mm 水泥砂浆；
　　60mm 混凝土垫层；素土夯实
9　φ50mm 镀锌钢管花架，做涂料
10　φ10mm 不锈钢缆及配件
11　种植土
12　灯具
13　φ8mm 不锈圆钢固定钢缆下端在混凝土锚
　　块上
14　钢筋混凝土墙外做防水层及保护层

Linear courtyard wall section

1　Galvanized steel railing, painted
2　2.5mm aluminum sheet cap
3　Roof construction:
　　Treated wood deck; 40mm reinforced
　　concrete topping; building paper; roofing felt;
　　15mm cement mortar; 40mm rigid insulation;
　　lightweight concrete to form drainage slope;
　　site-cast reinforced concrete slab
4　Lime plaster ceiling finish, painted
5　Steel beam with fireproofing coating, painted
6　Aluminum curtain wall, double glazing with
　　low-E coating
7　XPS insulation panel with gypsum board
　　facing, painted
8　Floor construction:
　　Ceramic floor tile; 20mm cement mortar;
　　40mm concrete topping; waterproof coating;
　　20mm cement mortar; 60mm concrete;
　　compacted soil
9　φ50mm galvanized steel tubing frame,
　　painted
10　φ10mm stainless steel cable and fittings
11　Soil
12　Lighting fixture
13　φ8mm stainless steel bar anchoring the
　　lower end of cable to concrete cube weight
14　Reinforced concrete wall with waterproofing
　　and protection layers

0　　　　　　　　　500mm

一楼由条形庭院隔开的小餐厅，向西望室外平台。
One of the first-floor dining spaces defined by linear courtyards, looking toward the deck.

穿过层层条形庭院南望，庭院中种植攀援型植物。
Looking south through layers of linear courtyards in which climbing plants are planted.

从二楼大餐厅望屋顶平台及 A 字形爬藤花架（下为条形庭院）。

From the large dining space on the second floor, looking toward the roof deck and the plant-climbing A-frames (above the linear courtyards).

建筑北立面。
North elevation of the building.

建筑局部采取园林的形态
rchitecture Partially Taking the Form of
andscape

III.

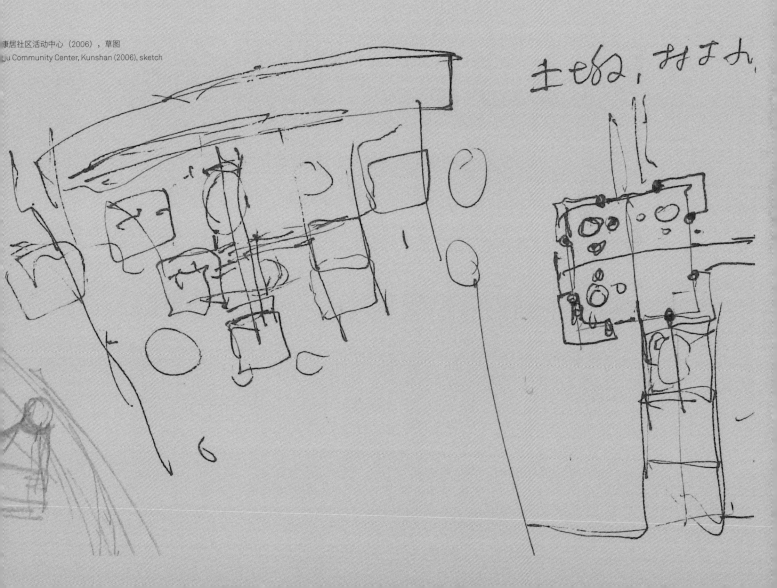

康居社区活动中心（2006），草图
ju Community Center, Kunshan (2006), sketch

"山脉"建筑
——昆山风尚公园多功能建筑群（方案，2007）

Architecture as Hills
—Mixed-Use Complex, Lifestyle Park, Kunshan (Scheme, 2007)

与其他江南水乡城市一样，历史古城昆山的中心城区（玉山镇）内有多条河道。到 2006 年，经过二十多年的改革开放，城市建筑的面貌已完全改变，但市区中的河流两岸并没有充分发挥其作为公共空间的作用。昆山市于此时委托其他设计单位做了一个环城滨江景观带方案，对从中心城区西边的叶荷河到北边的北环城河的水岸及七块河边空地做了新的用地规划及景观设计。其中最大的一块是位于北环城河中的河心岛，面积约为 3.78 公顷。岛北、西、南三面的河对岸为住宅、学校等功能的建成区；岛东隔水紧邻中心城区中最大的公共绿地，约 56.67 公顷的亭林公园。岛本身原是不对外开放的闲置林地，在新规划中将成为与河两岸用步行桥连接的风尚公园。与只有绿地的普通公园不同，这是一个集园林、展览、科普教育、娱乐、商业等为一体的综合休闲空间。为此在园中规划了多栋建筑（详见公园总体规划）。我承担了其中的多功能建筑群（A、B、C 栋）（已完成施工图）及另外四个独立单体（F、H、J、K 栋）的设计。以下主要介绍 A、B、C 栋的设计想法。

要吸引周边居民过桥越河来使用园中设施，建筑设计必须充分利用基地是市中心难得的一片绿洲这一独特性。建筑形式应设计成园林的合作伙伴，而不是市区中常见的广场上耸立的购物中心。我国江南城市有不少传统的公共休闲中心，像上海豫园、苏州虎丘、无锡天下第二泉等，均采用了将宗教、文化及商业建筑与自然或人工山水穿插结合的布置，最终形成的"景区"能全方位地满足游人的休闲需要，从祭奠、赏景、参观到餐饮、购物等。这一传统环境类型与现代单一功能的公园或商场大不一样，但仍受到今天公众的欢迎。借鉴它们的设计策略不仅呼应了前述的规划定位，也使风尚公园更能体现本地文化的特征，对在大规模城市更新中失去城市个性的昆山来说特别重要。本设计尝试了三个手法来实现建筑与园林的合作。

建筑局部成为"山脉"

实现上述合作的最大挑战是基地上的现有园林景观比较平淡。树木高度不大且彼此接近；地形为平缓土坡，只有东、中、西三个 2 到 5 米余高的小山坡，与河心岛东南侧亭林公园内 80.2 米高的玉峰山不可同日语。另一个挑战是为了容纳展览、娱乐、手工艺村等功能，本建筑群必须有足够大的建筑面积，有可能破坏与园林的平衡。

作为一个一石两鸟的解决办法，本设计将建筑群的大部（B、C 栋）分成"基座"与"上部建筑"两部分。一层高的条形基座（地面标高约 7.00 米以下）中的大部段落埋入现有小山坡中，露出部分的外表大部做成实墙（部分为斜面），材料为带条纹模板的清水混凝土或板岩，并覆盖攀援植物如爬山虎，使整个基座成为连接现有西、中部两个小山坡的人造"山脉"。在基座上另设几何形态截然不同的小块状上部建筑（地面标高约 7.00 米以上）。外立面以轻盈通透的精致材料为主，如玻璃、钢构架、木板或铝板墙面。这些位于二层的体块有的面对岛西河道中心，有的面对建筑东面的小广场，

与现有周边景观相呼应。

由于基座外形埋入土中或貌似山坡，基座与上部建筑之间在外表材料上又形成自然与人工的对比，这些都加强了基座作为山坡一部分的幻觉，增加了山坡的体量感、峭度，提升了园林景观在合作中的表现力。由于观众仅把上部建筑当成房子，从而获得建筑面积的同时减少了对现有绿化环境的影响，可谓两全其美。

必须强调的是，这一将建筑部分地变形为园林景观的手法是提示性的借鉴，从基本几何形态及材料特征上着眼，而非在房屋上虚假地仿制"自然"形态细部，如设计中的山坡仍然是规则的清水钢筋混凝土斜板。同时，被变形的建筑空间必须满足功能要求，如本设计中基座埋入土中的部分（B 栋）有一边全为 6 米宽的下沉庭院，这些面积均有自然通风采光。另外，基座中空间的地面标高仍比室外一层地面高，减少了排水问题隐患。

由于人脑依据遗传的心理模式来认知环境，而自然环境在人类的进化过程中对这些模式的形成影响最大，因此，能唤起对典型自然山水景观（像峭壁、岩洞、悬崖）联想的建筑形式将最有感染力。除了上述的山脉及山腹中的下沉水庭院外，我还在建筑群的其他几处用几何形态的建筑形体来暗示自然意象。如建筑群中部大门内的庭院（B、C 栋之间）隐喻了峡谷，A 栋的室外楼梯象征了山道等。经验还证明，由构筑物形成的"山"体，比自然土坡更能在不大的园林中形成陡峭感，苏州园林中用湖石山峰，即为同理。以上这些均说明本手法在满足功能以外还有助于创造新景观，从而弥补河心岛现有园林平淡的缺点。

建筑拥抱广场、小湖

风尚公园是市中心难得的一片新的公共空间。公园总体规划中为此在河心岛布置了一个面临小湖的广场，可做集会表演或周末市场用。根据建筑与园林合作策略，建筑方案将 A、B、C 栋设计成一个用天桥、平台等连接的序列，从南、北、西面环抱广场及小湖，而不是做成规划中的三个分散的孤零零的单体。这不仅方便人从一处漫游到下一处，而且促进了建筑与园林中人的互动。建筑群的连续基座面临广场的一边被处理成种树木的台阶及平台，可做观众席用。部分上部建筑及其连廊临广场，边缘处还设有"美人靠"（窗台座凳）。这些措施都有意地促使周边建筑内的游人观赏广场上的活动，或使广场上的人反过来眺望周边人流及建筑窗内的展品，形成"人看人"的交流。为点明此用意，设计还在广场中另设一自然形状的平台用作简单表演的露天舞台。

带绿化的台阶和平台不仅方便休息及观赏功能，它们的外观也在垂直的建筑与水平的广场之间形成更自然的过渡，并隐含"山麓"的意境。江南不少在山林中开设公共活动中心，如苏州虎丘、杭州西泠印社等，都应用了类似的手法。本设计希

对该传统的借鉴来点出本土文化的影响。

建筑与景观的环抱关系还改善了原规划的园林形式。通过将建筑群的一部分（C
边缘东段）悬挑到湖的北端，使原本突然终止的水面产生了在建筑下继续延伸到
城河的幻觉，扩大了整个园林环境的空间尺度。

外空间互动

为了强调建筑群位于园林之中的特点，本设计利用各种可能，创造了多种室外活
间，并使它们与室内出租面积相邻，有利于休闲活动溢出到园林中进行。具体的
不仅体现在一层室内与附近地面广场或庭院之间的关系，而且体现在二层商店旁
顶平台等。室外空间均享有优美景观（特别是各屋顶平台），其中大部分还设置
架或其他绿化以提供遮荫。这些室外空间可用作露天咖啡座或餐座、盆景展示、
品陈列等。这种室内外空间配对来服务于一种建筑功能的做法同时也是我国建筑
的优秀传统之一。(2007)

tract

ng crisscrossed by many rivers, Kunshan started in 2006 the redevelopment of its
erfronts to revitalize their role as public spaces. As part of the project, a vacant
island in the city center became the 3.78-hectare Lifestyle Park, which would
bine cultural and commercial uses in a landscaped setting. To make full use of
rare green site in the urban core, the scheme for the Mixed-Use Complex (Buildings
, C) integrates the building with the landscape. As its precedents, many historic
nic areas in the region become successful public leisure centers by interweaving
landscape with temples and shops.

site displays a plain landscape of three small hills, which could easily be
whelmed by the large building volume for the planned functions. To solve both
blems, the design divides the building into a "podium" and several "superstructures."
buried in the existing hills, the podium connects two hills together, amplifying
overall size and steepness of the new "hill." In contrast to the solid exterior of the
ium, the superstructures exhibit a lighter, more transparent form. Thus people see
ller "buildings" floating on top of the "hill," while the latter contains a large amount
oor area with natural lighting and ventilation supplied by sunken courtyards.
ughout the building complex, the design also creates several other spaces that
y landscape features such as canyon and mountain path.

einforce the integration between the buildings and the landscape, the scheme
es Buildings A, B, C a continued sequence embracing the planned plaza and lake.
ped platforms are created along the plaza-side of the podium, looking toward a
e in the plaza. The superstructures and the corridors connecting them all have
e windows toward the plaza. These measures encourage visual interactions
ween activities in the plaza/waterfront and in the surrounding buildings.

ng advantage of the landscaped site, the scheme creates many outdoor spaces
to indoor ones, such as the courtyards in the podium and the roof decks in the
erstructures. With beautiful views and good shading, these outdoor settings allow
r activities to spill out, recalling the traditional Chinese courtyard architecture.

工程资料

地点 江苏省昆山市亭林公园西面，北环城河中河心岛
时间 2007
建筑面积 4 820 平方米
业主 江苏省昆山城市建设投资发展有限公司
设计
建筑：缪朴设计工作室（缪朴）；上海天维建筑设计事务所
结构：上海源规建筑结构设计事务所（张业巍、李明蔚、杨振宇）
设备：上海源涛机电工程设计事务所
规划与园林：EDAW（香港）

Project Data

Location River Island in North Ring River, west of Tinglin Park, Kunshan, Jiangsu Province, China
Project Period 2007
Floor Area 4,820 square meters
Client Kunshan City Construction, Investment and Development Co., Ltd.
Designer
Architecture: Miao Design Studio (Design Architect), Pu Miao; Shanghai Tacon Design (Architect of Record)
Structure: Shanghai Yuangui Structural Design Inc., Zhang Yewei, Li Mingwei, Yang Zhenyu
Engineering: Shanghai Yuantao Mechanical and Electrical Engineering Design
Planning and Landscape: EDAW (Hong Kong)

N

0   20m

已设计建筑 BUILDING（DESIGN COMPLETED）

已规划建筑 BUILDING（PLANNED）

北环城河

NORTH RING RIVER

步行桥 FOOTBRIDGE

风尚公园
LIFESTYLE P

C

D

B

H

A

步行桥
FOOTBRIDGE

桥 BRIDGE

8.40

叶荷河
YEHEHE
RIVER

2.40

湿地 WETLAND
0.30

0.30

2.35

4.50

3.00

2.50

2.50

2.50

7.00

3.00

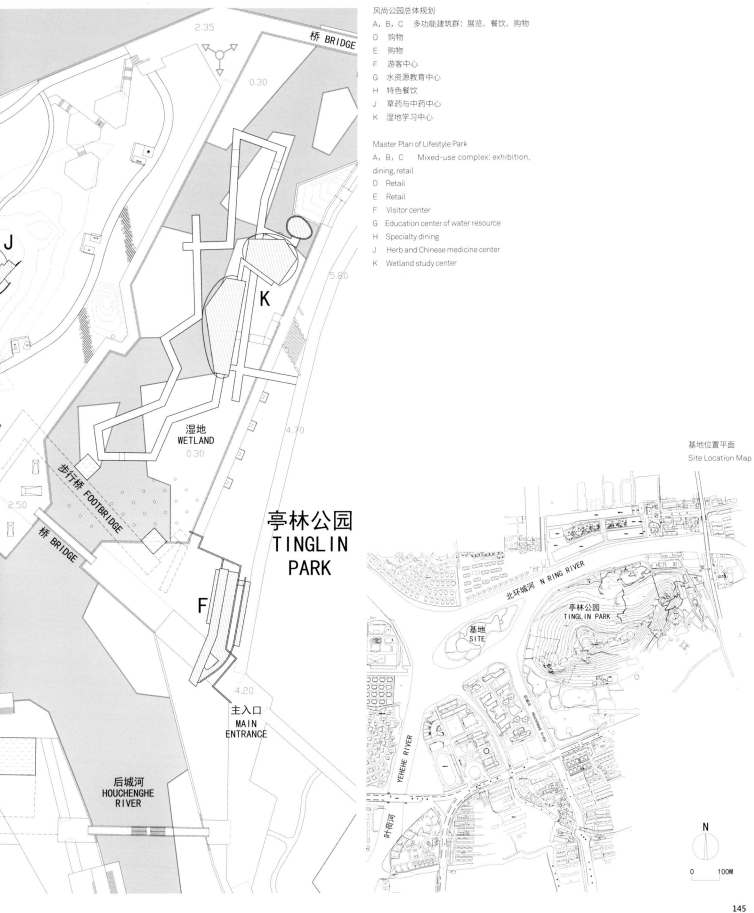

风尚公园总体规划

A，B，C　多功能建筑群：展览、餐饮、购物
D　　购物
E　　购物
F　　游客中心
G　　水资源教育中心
H　　特色餐饮
J　　草药与中药中心
K　　湿地学习中心

Master Plan of Lifestyle Park

A，B，C　Mixed-use complex: exhibition,
dining, retail
D　　Retail
E　　Retail
F　　Visitor center
G　　Education center of water resource
H　　Specialty dining
J　　Herb and Chinese medicine center
K　　Wetland study center

桥 BRIDGE

2.35

0.30

J

5.80

K

4.70

湿地
WETLAND
0.30

步行桥 FOOTBRIDGE

2.50

桥 BRIDGE

亭林公园
TINGLIN
PARK

F

4.20

主入口
MAIN
ENTRANCE

后城河
HOUCHENGHE
RIVER

基地位置平面
Site Location Map

北环城河 N RING RIVER

亭林公园
TINGLIN PARK

基地
SITE

YEHEHE RIVER

HOUCHENGHE RIVER

叶荷河

N

0　　100M

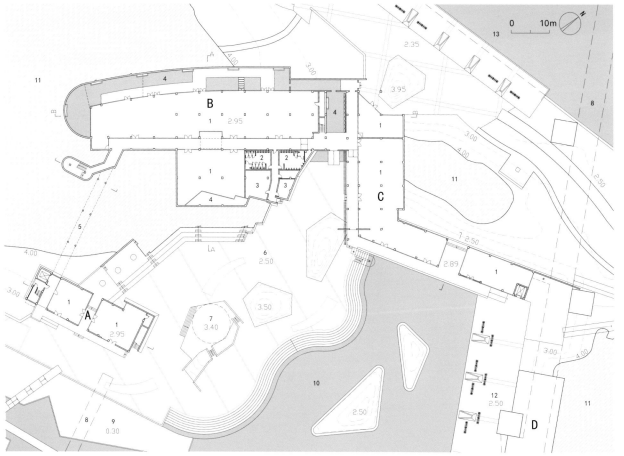

从东南面鸟瞰建筑群全景。
An aerial view of the southeast side the complex.

A、B、C 栋一层平面

1 出租
2 公厕
3 机房
4 庭院
5 天桥
6 广场
7 舞台
8 步行桥
9 湿地
10 小湖
11 现有山坡
12 平台
13 北环城河

Buildings A, B, C First Floor Plan

1 Rental
2 WC
3 Mechanical
4 Courtyard
5 Skybridge
6 Plaza
7 Stage
8 Footbridge
9 Wetland
10 Lake
11 Existing hill
12 Deck
13 North Ring River

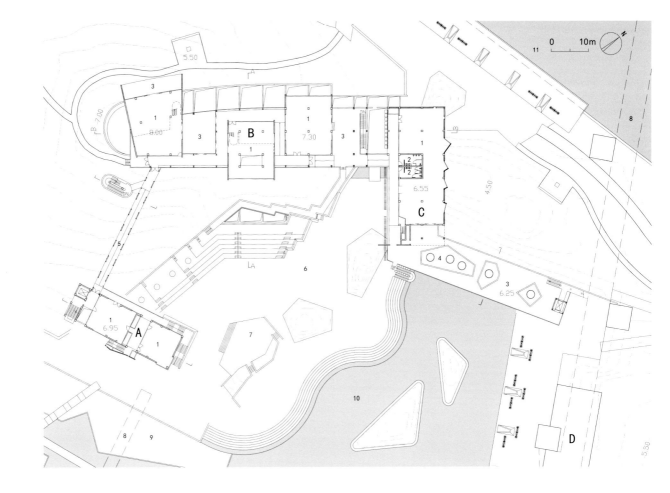

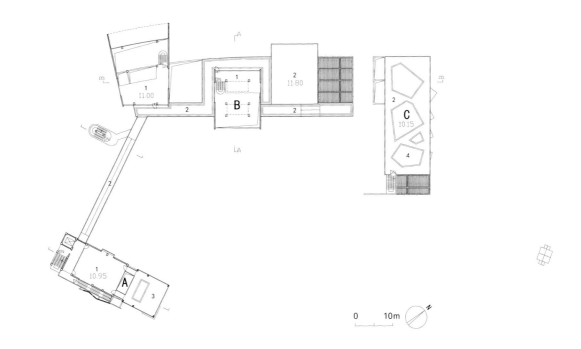

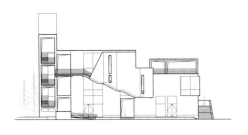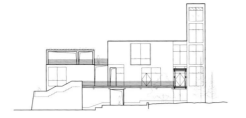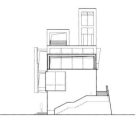

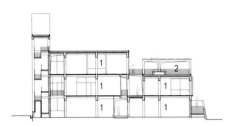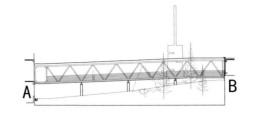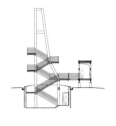

0　　　10m

A 栋：（上左）南立面，（上中）北立面，（上右）
东立面，（下左）剖面，（下中）天桥东立面，
（下右）天桥剖面
1　出租
2　平台

Building A: (upper left) South Elevation,
(upper middle) North Elevation, (upper right)
East Elevation, (lower left) Section, (lower
middle) Skybridge East Elevation, (lower right)
Skybridge Section
1　Rental
2　Deck

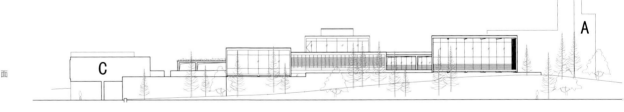

B 栋：（上）西立面，（下）A-A 剖面
1　出租
2　庭院
3　广场
4　舞台
5　小湖

0　　　10m

Building B: (upper) West Elevation, (lower) A-A
Section
1　Rental
2　Courtyard
3　Plaza
4　Stage
5　Lake

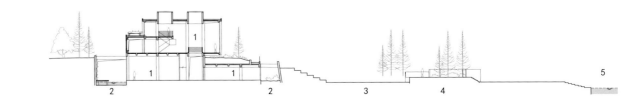

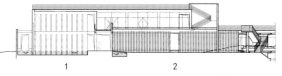
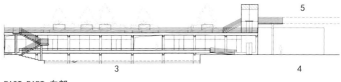

西部 WEST PART   EAST PART 东部

0   10m

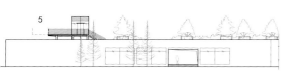

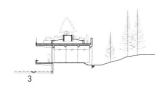

东部 EAST PART   WEST PART 西部

C 栋：（上）南立面，（下左）北立面，（下右）　　Building C: (upper) South Elevation, (lower left)
剖面　　　　　　　　　　　　　　　　　　　　　　North Elevation, (lower right) Section
1　庭院　　　　　　　　　　　　　　　　　　　　1　Courtyard
2　广场　　　　　　　　　　　　　　　　　　　　2　Plaza
3　小湖　　　　　　　　　　　　　　　　　　　　3　Lake
4　平台　　　　　　　　　　　　　　　　　　　　4　Deck
5　步行桥　　　　　　　　　　　　　　　　　　　5　Footbridge

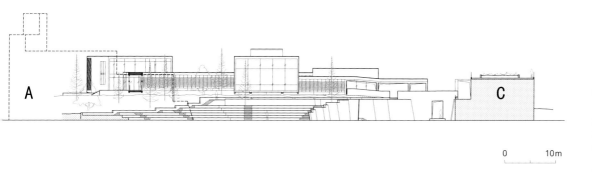

0   10m

B 栋：（上）东立面，（下）B-B 剖面
1　出租
2　平台
3　庭院
4　种植坛

Building B: (upper) East Elevation, (lower) B-B
Section
1　Rental
2　Deck
3　Courtyard
4　Planter

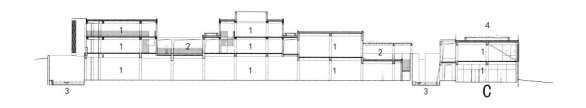

从广场舞台上望 B 栋东面，建筑群中部大门（B、C 栋之间）及 C 栋南面。

From the stage in the plaza looking to the east side of Building B, the middle entrance of the complex (between Buildings B and C), and the south side of Building C.

隔湖望建筑群东面外观。A、B、C 栋分列左、中、

Looking across the lake to the east sid
Buildings A, B, C (left, middle and right in the pic
respectively).

从湖中望广场及建筑群东北面。
From the lake looking into the plaza and the northeast side of the complex.

筑群中部大门望 B 栋（左）、C 栋（右）之间
水庭。
n the middle entrance of the complex looking
the water courtyard between Buildings B (left)
C (right).

B 栋一层室内景观及西、南边室外水庭。
The interior of the first floor in Building B and the water courtyard along the west and south sides.

从北环城河桥上看建筑群北面外观。
From the bridge on the North Ring River looking to the north side of the complex.

从北环城河中望建筑群西北面全景。
From the North Ring River looking to the northwest side of the complex.

（右页上）B栋西面外观。
(Right upper) The west side of Building B.
（右页下）B栋（左）、天桥（中）及A栋（
西南面外观。
(Right lower) The southwest side of Building B (
Skybridge (middle), and Building A (right).

153

"丛林"建筑
——昆山康居社区活动中心（2006）

A "Forest-Like" Building
—Kangju Community Center, Kunshan (2006)

近年来国际上流行将整个建筑包在一个简单几何形中的做法，比如像在所有立面上覆盖一层同质的"膜"或将建筑外形做成曲面的"blob"。正如其他西方时髦风格一样，国内对此也出现了一批模仿者。这种风格当然是建筑师可以选择的处理手法之一，但我们在使用前应当先了解它与生俱来的问题。比如说，整个建筑必然呈现为一个尺度超大但颜色、材料单一的实体，给人行道上的使用者以威严压逼的感觉；其次，覆盖建筑全部外表的"膜"（或网、幕墙等）为了维护自己的同质与完整，必然要限制门、洞等的开启，从而拒绝了人在建筑与城市环境之间的流动以及在建筑中同时使用室内外空间的可能；第三，这种手法最好是用在极少数城市地标类型的项目上，而且周围必须有许多缓冲空间，否则就会产生因与贴近的城市文脉争斗而两败俱伤的感觉。顺便说一句，这一风格显然是两个西方建筑传统在今天的延续：将室内外空间截然分开的"绿地环绕城堡"模式，以及将建筑外貌处理成主要由面组成的"雕塑"模式。

Blob 之类在近年中的流行，究其深层原因，还是由于今天的前卫建筑师们日益背离了早期"现代建筑"的基本原则，把自己的工作等同于变幻表面风格。而在西方后现代资本主义社会中，新颖风格所创造的商品利润份额日益增大，这又使前卫建筑师们急欲垄断这个市场。其所用的经营策略是建筑师明星化；设计策略则为自我孤立，以达到强调品牌独特性的目的。比如有意地制造出一些不太适用的，给人以冰冷、敌意以至危险感的环境形式，以提醒消费者他们所购买的是不必遵守日用商品规则的"奢侈品"。针对国内对这些"酷"风格的吹捧，我们不禁要问，难道人的日常感情，比如人在建筑中普遍期望的安定感与庇护感"过时"了吗？我与合作者在最近几年中做了一系列实验，其目的就在于探索另一条路，一条主要是通过感动而不是震动人的创新之路。以下介绍的昆山康居社区活动中心是其中一例。

在江苏省昆山市的西郊，一个占地 80 余公顷、人口 3 万的康居小区正逐步成形，将主要用来安置因城市改造而迁移的老城居民。在小区的中心地带规划了商业设施及一个社区公园。康居公园占地约 3.3 公顷，西边与商场为邻，东、北两面与住宅区相通。在公园中心的小广场北面规划了一个建筑面积约 420 平方米的社区活动中心，主要用作茶室及各种小型社团的活动。建筑周围现有的景观包括小广场东侧的半圆形水池，以及从公园南面主入口延伸过来的一片湿地及一条木栈道。

人在休息时最喜欢自己能受到保护，但同时又可以在最大程度上观察其他人或空间。英国景观理论家杰伊·阿普尔顿（Jay Appleton）在他的《风景的体验》（1975）一书中对这个人人都能领会但又说不清的感觉做了深入的分析。阿普尔顿提倡的有"视野"的"庇护所"（Prospect-refuge）是什么样的呢？首先，这一空间的外表应是布满孔洞空隙的，暗示人到处都可以进入藏身，同时也可随时自内向外观望。康居社区中心位于公园正中，又是园中唯一一栋建筑，很自然地成为从四方过来的人流的歇脚之地。这决定了该建筑的各个面都应具有"丛林"或"蜂窝"状的外观，以显示对居

民的接纳欢迎。本中心因此被设计成一个由多个独立或半独立单体组成的建筑群，体之间形成多种空隙或洞口。这与本文开首所说的用一张"膜"来隔断建筑与周围环境交流的时髦做法正好相反。

康居社区中心的使用者将是三五成群的附近居民，来这里打牌、下棋、聊天。一个群体以至个人都希望能在建筑内找到自己的"庇护所"。经验告诉我们，有情感的室内空间总是含有许多可让人们凭靠的边界和角落，从而创造出似隔不隔的空间，满足人在公共场所中希望为自己划出一个领域、又能窥视他人活动的双重愿望。上海城隍庙湖心亭茶室（清末民初）含有多重转折的外墙并将窗台做成可坐的"美人靠"，就是为了创造更多靠边角座位的佳例。本中心由多个 L 或锯齿形状单体组成，从而在一个小建筑中产生了二十多个转角，增加了外围边界的长度。沿大部分外墙边缘设计了像传统"美人靠"的窗台座凳，与室内可移动的桌椅形成一个个四到六人的组团。组团所产生的隔断或半隔断空间成了大量小团体可占领的"房间"。

这一组团概念同时反映在建筑不同寻常的结构布局上。我们采用了比常规较大的 3300 毫米 ×3300 毫米柱网，每一网格可容纳四个组团，网格边的柱子有助于在空间界定人的社交空间。由于这些组团空间各自占据网格的四个角，我们特意将柱子从网格的四角移向每边的中点，使坐在每个角落两边的使用者没有被柱子分开的感觉。玻璃窗在这些角落的转折边缘上也采用了无框设计。

为了创造让人可以观察探索的"视野"，社区中心的所有房间都开向多种室内外景观。其中不少房间面对建筑西南面的湿地、东南面的半圆形水池及其他公园中的绿化。本工程的部分屋顶还被设计成绿化屋顶，为二层面向对应角度的房间提供了景观。我们在建筑形体组合上还有意创造了三个内庭院，人们必须通过狭窄通道或从天桥过街楼下面穿过才能进入。同时，建筑群内各个单体的外墙均为布满密集窗框的玻璃幕墙，部分玻璃并做成半透明。通过以上的空间布局及界面处理，产生了一些半明半显的"悬念"空间，使人们无论是在建筑内外，都可以不时透过一个尺度不大的窗口或重重互相局部遮掩的界面，看到远处某个院落或房间的一角，从而在使用者心中引起对看不到部分的神秘感。

除了上述的中心设计概念以外，我们还继续试验了将室内外空间配对服务于同一主要建筑功能的概念。在近距离内可以同时得到室内环境的方便舒适，又能享受室外微风花影的自然气氛，是我国传统建筑在创造宜人环境上的宝贵遗产之一，四合院就是一个典型的例子。可能是由于文化及气候的原因，西欧建筑传统从来就忽视这种空间布局，而更倾向于将基地上的所有室内空间集中成一个被绿地环绕的实体。因此，如何使用现代技术在多层建筑及高密度城市中重现室内外空间配对，是现代建筑中国本土化的一个重要课题。在本工程中，我们除了在地面布置了多个庭院外，更在二、三层的屋顶设计了带遮阳花架的屋顶平台。无论是在哪个茶室中，人们都

室内移座到附近的室外环境中去，实际上就是在三维空间中应用了四合院的概念。

门还注意到在我国传统风景区中，经常有一些不是全封闭的室内空间，如亭榭、岩□等。由于与室外联系方便，它们成为最受欢迎的公共空间。我们为此在本工程的南□中部结合下沉庭院设计了一处类似的半室外空间，在里面设置了刻有棋盘的石桌凳。

在立面设计上，我们试图在建筑的实体外表上体现一种所谓江南文化的精致与凝□。这种感觉主要是通过三种尺度的线性构图（而非西方传统中的面构图）来实现的。□飞，深灰色的钢筋混凝土梁柱勾画出了大尺度的基本框架，为此我们还试验了染色□水混凝土的新工艺。其次，黑色的铝窗框、遮阳百页、雨水管及本色清漆木花架条□供了中间尺度的层次。铝窗的分割被有意向人体尺寸接近，因为我们认为"墙"的□本意义是为人提供保护，在需要保护的地方使用超大尺寸的玻璃是藐视人在建筑中□基本期望，更不用说这样的幕墙不允许室内外的交流。最后，纤细的黑色钢栏杆以□不锈钢爬藤钢缆补充了细节上的层次。（2010）

□stract

□ 3.3-hectare Kangju Park is encircled by apartments of a new residential district.
□he heart of the park Kangju Community Center was planned for surrounding
□dents to enjoy tea, play cards, and conduct other social activities in small groups.

□axing in a public space, people love to be in a "refuge" with a "prospect." Such
□ace often has a porous exterior to welcome entering and to allow looking out.
□refore the community center is designed as a forest-like structure composed of
□tiple small building volumes with gaps among them. Inside, this configuration
□ates numerous boundaries and corners to form small territories without solid
□ls. Columns are moved away from the corners to the middle of a building volume,
□phasizing the corner as a unified place.

□rovide many "prospects," every room opens up toward one or more exterior views
□ch include the landscape in the park and the courtyards and roof gardens in the
□ding. The forest-like spatial structure also creates half-revealed "suspenseful"
□ces, allowing people to frequently see through a small opening a partial view of a
□ant courtyard or room.

□overall spatial layout juxtaposes rooms with courtyards and roof decks. People
□easily move from an indoor setting to exterior seats under the green trellises. Such
□iring between indoor and outdoor spaces is a tradition in Chinese architecture.
□ese traditional scenic areas often contain semi-outdoor public spaces, like
□lions and caves, which are loved by the public. This design also creates one such
□ce in the middle of the south side.

资料

江苏省昆山市前进西路（思常路口）

2005—2006

□积 420 平方米

江苏省昆山城市建设投资发展有限公司

缪朴（上海市园林设计院顾问）

庄伟（上海市园林设计院，下同）

许曼

□水：李雯

周乐燕

□mus》（中文版，2008 年第 4 期），《新建筑》（2011 年第 3 期）。

□'07'06'05——当代中国建筑创作论坛作品集》（2007），《当代中国建筑地图》（2014）。

Project Data

Location West Qianjin Road at Sichang Road, Kunshan, Jiangsu Province, China
Project Period 2005-2006
Floor Area 420 square meters
Client Kunshan City Construction, Investment and Development Co., Ltd.

Designer

Architecture: Pu Miao (Design Architect), Shanghai Landscape Architecture Design Institute (Architect of Record)
Landscape: Zhuang Wei (Shanghai Landscape Architecture Design Institute, same below)
Structure: Xu Man
Plumbing: Li Wen
Electrical Engineering: Zhou Leyan

Publication

Domus (Chinese Edition, 4/2008), *New Architecture* (3/2011).
Architecture'07'06'05: Works from the Forum on Chinese Contemporary Architecture (2007), *Atlas of Contemporary Chinese Architecture* (2014).

总平面		Site Plan	
1	社区活动中心	1	Community center
2	凉亭	2	Pavilion
3	空调设备	3	AC equipment
4	林荫广场	4	Plaza with trees
5	水池	5	Pond
6	湿地	6	Marsh
7	木栈道	7	Boardwalk
8	商场	8	Retail
9	公寓	9	Apartments
10	前进西路	10	West Qianjin Road

0 30m

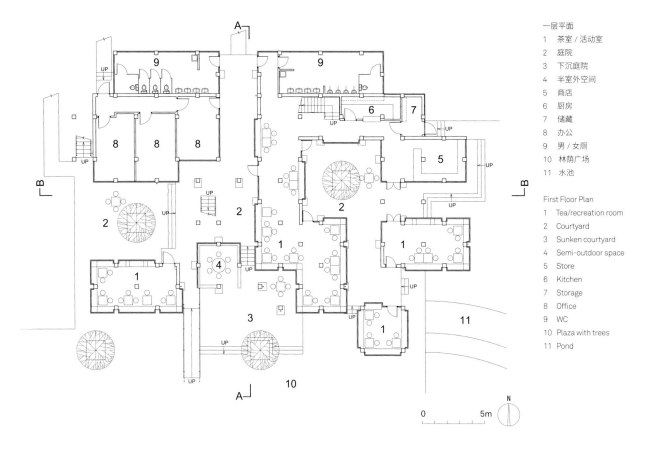

一层平面
1 茶室 / 活动室
2 庭院
3 下沉庭院
4 半室外空间
5 商店
6 厨房
7 储藏
8 办公
9 男 / 女厕
10 林荫广场
11 水池

First Floor Plan
1 Tea/recreation room
2 Courtyard
3 Sunken courtyard
4 Semi-outdoor space
5 Store
6 Kitchen
7 Storage
8 Office
9 WC
10 Plaza with trees
11 Pond

0   5m

N

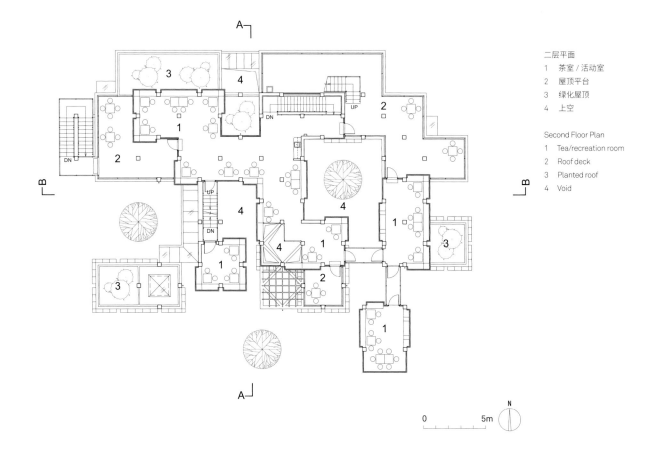

二层平面
1 茶室 / 活动室
2 屋顶平台
3 绿化屋顶
4 上空

Second Floor Plan
1 Tea/recreation room
2 Roof deck
3 Planted roof
4 Void

0   5m

N

三层平面
1 茶室 / 活动室
2 屋顶平台
3 天桥
4 上空
5 屋顶

Third Floor Plan
1 Tea/recreation room
2 Roof deck
3 Bridge
4 Void
5 Roof

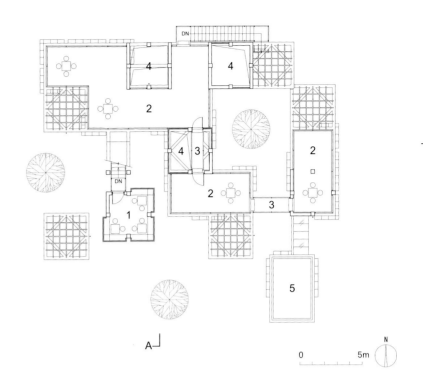

0　　　　5m

（上）A-A 剖面，
（下）B-B 剖面
1 茶室 / 活动室
2 庭院
3 屋顶平台

(Upper) A-A Section,
(lower) B-B Section
1 Tea/recreation room
2 Courtyard
3 Roof deck

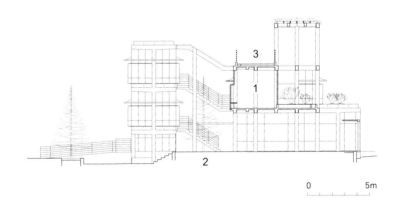

0　　　　5m

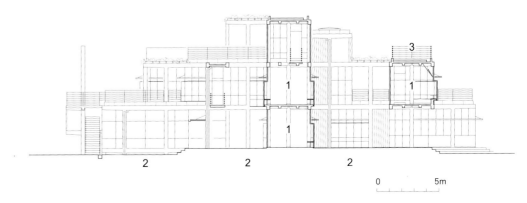

0　　　　5m

建筑群南立面中部。
Middle part of south elevation.

立面局部及一层的半室外公共空间。
se-up of south elevation and
semi-outdoor public space on
first floor.

从内庭院之一南望建筑外广场。
From one of the courtyards looking south toward the plaza outside of the building.

过建筑南侧的空隙望内庭院之一。

e of the courtyards seen through a gap in the
th elevation.

从单体之间的天桥下望内庭院之一。
One of the courtyards behind the bridge between
building volumes.

建筑东南角及公园半圆形水池。
Southeast corner of the building near the
half-circle pond in the park.

东立面。
East elevation.

建筑西边的屋顶平台及远处的屋顶绿化。
The roof deck on the west side of the building, with
a planted roof in distance.

室内二层的一个组团空间及"美人靠"。
One of the corner spaces for small groups (with window seats) on the second floor.

三层高室内拔高空间。
Vertical spatial penetration up to three stories high.

室内一层的一个组团空间及"美人靠"。
One of the corner spaces for small groups (with
window seats) on the first floor.

园林作为建筑中的城市公共空间
Gardens as Urban Public Spaces in a Building

IV.

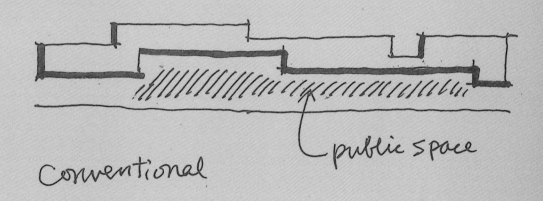

conventional

public space

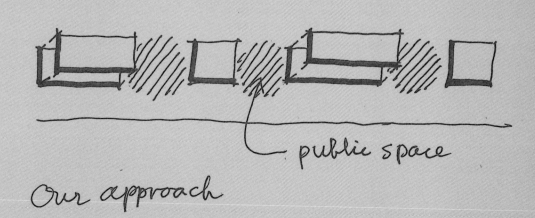

public space

Our approach

金谷园多功能建筑群（2013），概念示意
uyuan Mixed-Use Complex,
shan (2013), conceptual diagram

以公共园林为标志的社区中心
——昆山里库袖珍公园及社区中心（方案，2008）

Public Green as the Landmark of a Community Center
—Lishe Pocket Park and Community Center, Kunshan (Scheme, 2008)

我国绝大部分具有一定历史的城市都含有一个俗称"老城区"的市中心。其特点是人口与建筑稠密，公共空间总量少。除了少数在近代受西方规划思想影响的特例外，公共室外空间基本上都采取街道的形式，缺少像广场、公园那样的节点状空间。这些特点在今天产生的问题，首先自然是无法满足城市居民使用公共空间的需要，特别缺乏的是有利闲谈、聚会等向心型社交活动的节点空间。但还有另一个容易被人忽视的问题，由于人在城市中体验到的除了街道还是街道，没有节点空间来产生地标，这很容易使人产生视觉和行为上的单调感。这就使得老城区的城市环境缺少可识别性，也失去了为不同社区创造个性象征的机会。这一城市形态的缺陷是无法靠建筑立面上的变化来补救的[1]。

历史城市昆山的中心城区（玉山镇）就是这样一个典型例子。该区仅户籍人口密度每平方公里就至少有 1 550 人（2000 年）。区内满是林立的多层建筑拥挤在 5~7 米宽的街道巷弄两旁。可能是察觉到高密度城市的问题无法靠小动作改善，该镇的里库社区于 2007 年以高昂成本在里库新村靠北面县后街的一边换了原来的两栋住宅楼，取得一块约 20 米 ×55 米（1 097 平方米）的基地，要在这里建一个社区中心，内含社区居民委员会办公室、一个社区活动室及为邻近住宅楼服务的自行车库。但这只是任务书的一半，业主还要在同一块地上建一个小公园。由于基地旁的街对过是一个小学的大门，该绿地不仅将满足居民的日常休闲，还可让每天接送孩子上下课的家长有地方可坐。最后，公园又应成为里库社区的标志，比如说要设置两座纪念在本地出生的唐代诗人孟郊（751—814）的雕像。

如何在这样小的基地上满足如此多且不同的用途呢？业主请其他人设计的解决办法是一栋房子旁边加一块大大缩小的边角绿地。我认为这种只将不同建筑功能肩并肩地布置在同一个基地平面上的做法，已经脱离了今天高密度城市的现实，无法充分利用这样一块用高价在稠密住宅区中挖出来的难得的公共空间。

我的方案因此提出一个"垂直功能分区"的概念。首先将一至二层高的建筑基本铺满基地，容纳任务书要求的各种建筑室内空间，同时将上述整个建筑体的屋顶设计成公共绿地，实际上就是对基地做到"一地两用"：一至二层的室内主要为社区内居民服务，从南面的里库新村内进入；屋顶公园为城市服务，从西、北面的城市街道进入。由于屋顶公园占据整个基地，再加上对建筑高度的严格控制，保证了本工程的城市外观是一大片长方形的开敞空间，与基地周围五六层高的住宅楼群及线形的街道空间均形成鲜明对比。该方案在城市规划的层面上满足了用袖珍公园来疏解本街区建筑单调稠密的要求，并为里库社区树立了一个显著的地标。

实际上，我国许多高密度的新老城市虽然没有太多较大的公园或广场，但不少公共建筑（如地方政府大楼、公共文化设施、社区服务中心等）均各自带有一些入口广场、院落或其他小块附属绿地，只是由于传统的城市规划习惯将这些设施在城市中分

散布置，再加上仅有的几个公园通常被围墙环绕，使这些公共绿地无法聚合成一定总量来被人感觉到。如果在今后的城市规划及行政管理上能够允许将两三个公共设（如能含一个邻里公园则更好）成团布置，就可以将其各自的室外空间集中设计为片足够显眼的公共绿地。该绿地与城市街道之间无墙分隔，应可从城市街道及组团的公共设施直接进入。这一共享绿地既为这些公共设施的使用者提供了延伸部分功能（如等候、社区集会等）到室外的可能，也为周边居民增加了公共休息空间，时又为城市创造了可识别的节点地标。对我国的高密度城市来说，这一依托公共绿建设的多功能社区中心是一种新的建筑类型。它更符合我国居民对带园林建筑的偏能更有效地使用城市土地，也更能体现政府服务于居民的现代社会关系。本方案可被看作这一探索的一个微型尝试。

除了上述有关基本布局的规划主题外，本方案还在建筑具体设计的层面上做了下考虑[2]：

（1）如上文已说到，与公共绿地相邻的建筑室内功能，应与室外空间有互动联最好能允许室内活动在需要时可以延伸到园林中。本方案为此在建筑体中穿插了两庭院，与室内用大面积落地玻璃门窗分隔。在屋顶公园中休息的居民与社区活动室居委会的室内活动均可通过庭院产生斜向视觉交流，如可以观看活动室中老年居民练歌舞。活动室的使用者还可走入庭院中享受绿化。庭院还使所有建筑空间（含半下层）均有自然采光及通风。

（2）公共空间必须靠近城市人流才会有人使用。本方案沿街道设计了多个出入方便公众进入公园，同时还为残疾人设计了可到达公园中心部分的坡道。本设计体布局带来的一个挑战是如何缩小屋顶公园的入口与城市人行道的高差。为此将建筑体做成台阶形，分为从西向东逐步升高的三段。入口处的西段被处理为半地下自行车库，从而保证了其屋顶上的公园入口与人行道之间相差只有 1.35 米。这一又被做成两级 450 毫米高的树坛，使公园看去更像街道边自然隆起的小山。人行道上的树坛矮墙均设计为可坐，既方便接送小学生的家长等候，也有助将人流引入园。

（3）面对高密度人口的使用，城市绿地应以硬地为主。本设计排除了易于磨损不便保养的草坪水池之类形式，设计了可容纳大量晨练、社交活动的硬地。其中位公园中段的小广场可供小型社区演出、交际舞等小团体活动。

（4）绿化应主要采取"占天不占地"的形式。本公园中绿化的主要形式为树坛同时提供了花架等其他空中绿化。在两个庭院中及基地西边主入口台阶处设计了钢管框架，内挂上爬攀援植物的不锈钢缆。沿街道各立面中的墙面也将爬满爬山虎垂直绿化。所以本设计保证了相当高的绿化覆盖率，但仍旧提供了大量可为居民活的公共场地。

（5）注意冬天日照及夏天遮荫（后者对中国城市特别重要）。本设计中大量

叶树冠覆盖上述的硬地，再加上多个凉亭及花架等，将充分满足此项功能。

（6）应提供大量可坐设施。屋顶公园中众多的树坛、花坛、凉亭及跌落平台的边界，被设计为可坐矮墙或长凳。

（7）应提供饮食及文化设施。由于基地西边的临街商铺中已有食品零售服务，设计中不重复提供。本设计中将孟郊雕塑及背景墙组合设置在屋顶公园的东段，位整个游园路径的最高点。这一布置为本公共绿地增添了历史意义及特定地点的个性。

建筑外表材料主要包括以下几种：建筑主体为清水混凝土框架及墙；凉亭、花架轻型结构为深灰色钢框架；深灰色钢栏杆及栅栏；本色清漆木花架条、木座凳等。2008)

释

Pu Miao, ed., *Public Places in Asia Pacific Cities: Current Issues and Strategies* (Dordrecht, The Netherlands: Kluwer Academic Publishers, 2001), Chapter 13 "Design with High Density：A Chinese Perspective," pp. 274-275.

此地讨论的七项功能要求中，除第一条是特别针对本文提议的以公共园林为标志的多功能社区中心外，其他六条均为对城市公共空间及高密度城市中公园的一般要求。源自国内外大量调查结果及建筑师自己的研究，详见：William Whyte, *The Social Life of Small Urban Spaces*（Washington, DC: Conservation Foundation, 1980）；缪朴，《谁的城市？图说新城市空间三病》，《时代建筑》2007年第1期，第6页；缪朴，《"人造自然"还是"绿色大厅"——城市公园设计模式质疑》，《建筑师》第91期(1999年12月)，第40-49页。

bstract

e historic cores of old Chinese cities are characterized by their high density and lack nodal public spaces such as squares and parks. Apart from the shortage of social aces, the monotonous streets also present limitations to create landmark spaces a community's identity, making the city less legible.

2007, the Lishe Community in the central city of Kunshan planned to build a mmunity center and a pocket park on a site of 20 by 55 meters. Instead of a building

next to a tiny garden on leftover land, this scheme proposes a "vertical zoning" that covers the whole site with a building of 1-2 stories high and creates a park on top of the entire building roof, producing a noticeable urban void (a landmark space) amid the dense apartments of 5-6 stories high.

The building is stepped to reduce the height difference between the western park entries and the sidewalk. Park users can see the activities in the community center below through two courtyards. Considering the intensive use of the ground, the park has mostly overhead/vertical planting and hard paving, including a small plaza for group dancing. Many sitting places are provided, especially at the northwest corner where parents wait for their kids from an elementary school across the street.

The design suggests a new planning model that groups several public facilities together so their piecemeal open spaces can be combined to create a sizeable public garden, which will also act as a landmark in a dense city.

工程资料
地点 江苏省昆山市县后街 (里库路口)
时间 2008
建筑面积 900 平方米
业主 昆山市玉山镇里库社区
设计 缪朴设计工作室（缪朴）

Project Data
Location Xianhou Street (at Lishe Road), Kunshan, Jiangsu Province, China
Project Period 2008
Floor Area 900 square meters
Client Lishe Community, Yushan Town, Kunshan
Designer Miao Design Studio, Pu Miao

总平面
1　袖珍公园及社区中心
2　住宅小区
3　零售
4　小学
5　县后街
6　里库路
7　小区路

Site Plan
1　Pocket park & community center
2　Housing estate
3　Retail
4　Elementary school
5　Xianhou Street
6　Lishe Road
7　Estate road

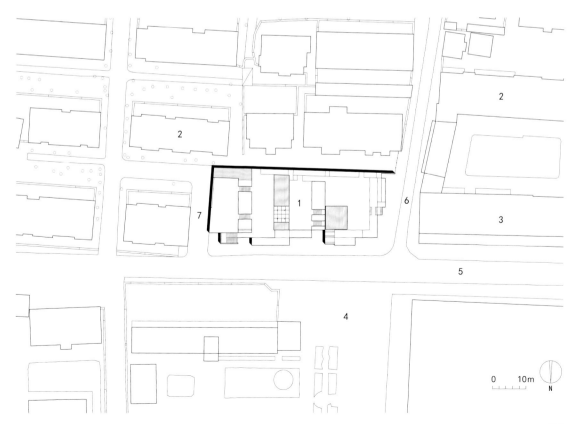

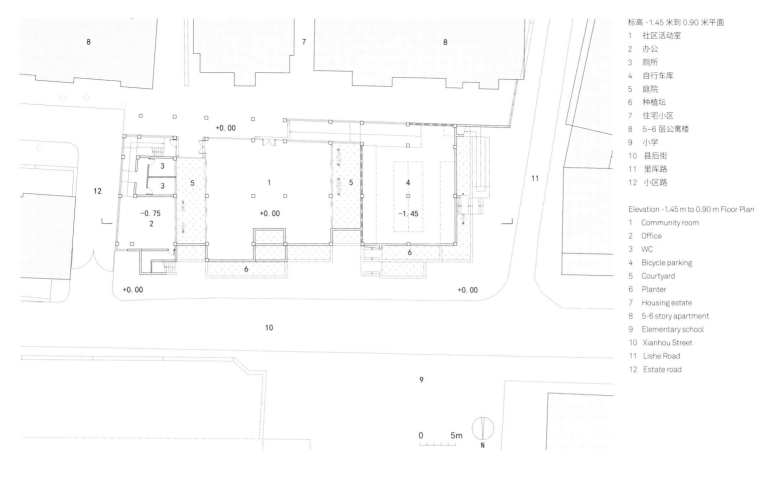

标高 -1.45 米到 0.90 米平面

1 社区活动室
2 办公
3 厕所
4 自行车库
5 庭院
6 种植坛
7 住宅小区
8 5~6 层公寓楼
9 小学
10 县后街
11 里库路
12 小区路

Elevation -1.45 m to 0.90 m Floor Plan

1 Community room
2 Office
3 WC
4 Bicycle parking
5 Courtyard
6 Planter
7 Housing estate
8 5-6 story apartment
9 Elementary school
10 Xianhou Street
11 Lishe Road
12 Estate road

0   5m

N

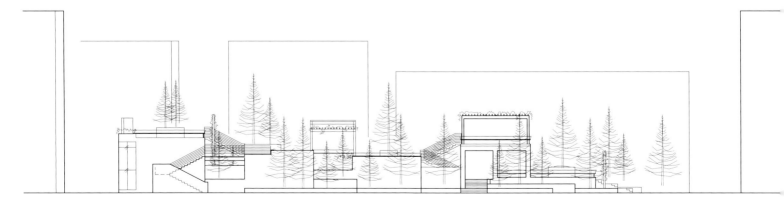

北立面
North Elevation

0   5m

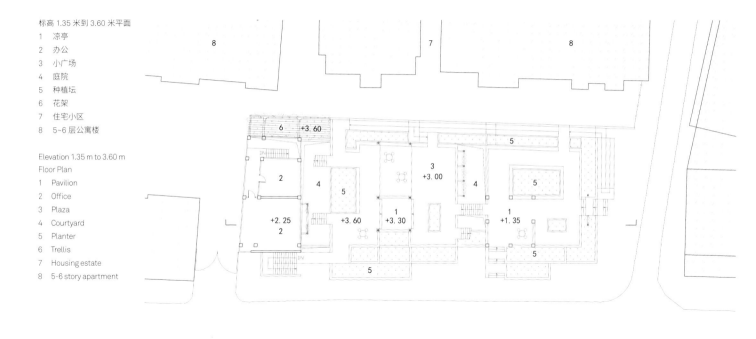

标高 1.35 米到 3.60 米平面

1 凉亭
2 办公
3 小广场
4 庭院
5 种植坛
6 花架
7 住宅小区
8 5~6 层公寓楼

Elevation 1.35 m to 3.60 m
Floor Plan

1 Pavilion
2 Office
3 Plaza
4 Courtyard
5 Planter
6 Trellis
7 Housing estate
8 5-6 story apartment

剖面 | Section

剖面		Section	
1	社区活动室	1	Community room
2	办公	2	Office
3	自行车库	3	Bicycle parking
4	凉亭	4	Pavilion
5	庭院	5	Courtyard
6	种植坛	6	Planter
7	5~6 层公寓楼	7	5-6 story apartment
8	里库路	8	Lishe Road
9	小区路	9	Estate road

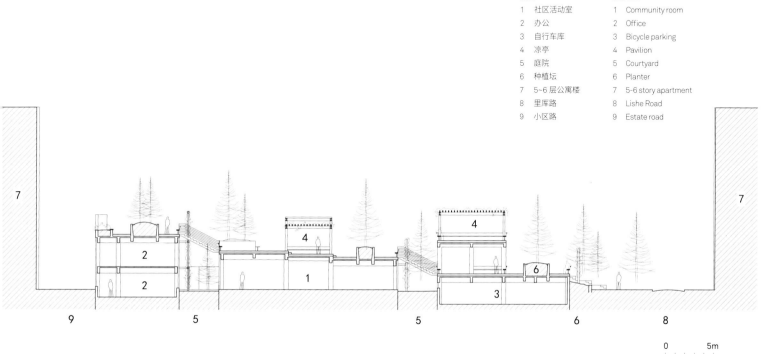

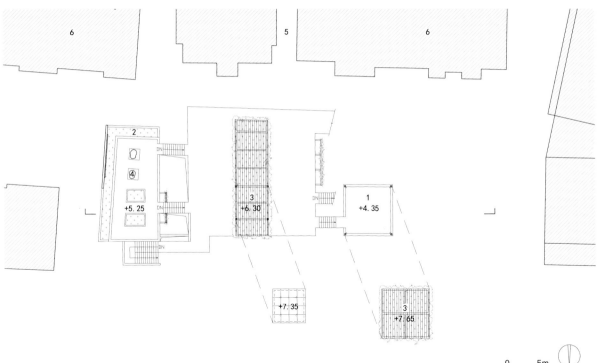

标高 4.35 米到 7.65 米平面
1 凉亭
2 种植坛
3 花架
4 雕塑
5 住宅小区
6 5~6 层公寓楼

Elevation 4.35 m to 7.65 m Floor Plan
1 Pavilion
2 Planter
3 Trellis
4 Sculpture
5 Housing estate
6 5-6 story apartment

6   5   6

2

4

+5.25

3
+6.30

1
+4.35

+7.35

3
+7.65

0   5m

N

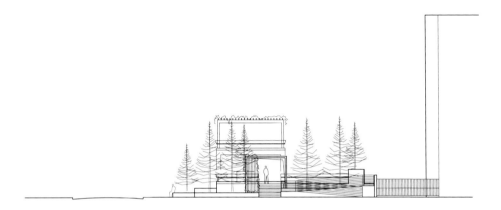

0   5m

西立面
West Elevation

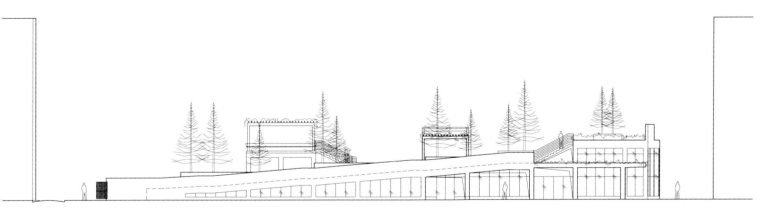

0   5m

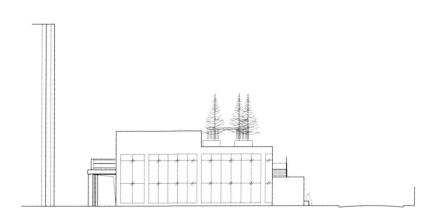

东立面
East Elevation

0   5m

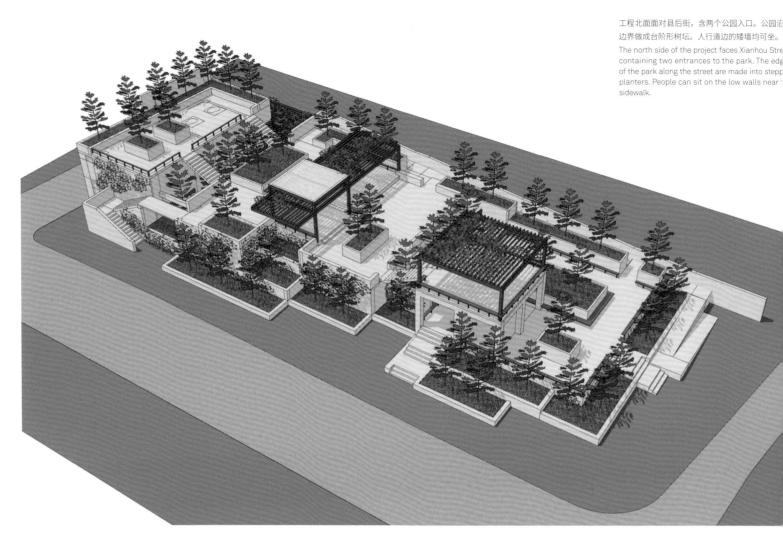

工程北面面对县后街，含两个公园入口。公园沿
边界做成台阶形树坛。人行道边的矮墙均可坐。
The north side of the project faces Xianhou Stre
containing two entrances to the park. The edg
of the park along the street are made into stepp
planters. People can sit on the low walls near
sidewalk.

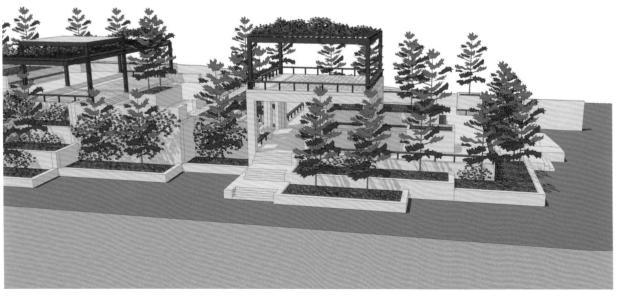

屋顶公园在工程北面的一个入口。
One entrance to the roof garden on the r
side of the project.

工程西面与里库路相邻，含两个公园入口。西北角
为接送小学生的家长设计了可坐矮墙。

The west side of the project is next to Lishe Road,
including two park entrances. Parents picking up
their elementary-school kids can sit on the low
walls at the northwest corner of the park.

築南面面向小区内部。居委会、社区活动室及自
车库的入口罗列在建筑右部。

th side of the project faces the interior of the
ate. The doors to the residents committee,
mmunity room, and bicycle parking are at the
t part of the building.

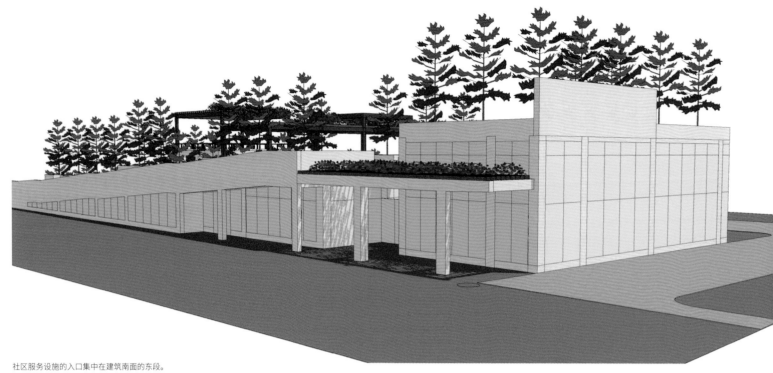

社区服务设施的入口集中在建筑南面的东段。

Entrances to the community facilities are grouped
in the east part of the south elevation.

工程东面面向小区内道路。东端二层建筑内为居委
会办公。

The east end of the project faces a road in the estate.
The two-story east building contains the residents
committee offices.

东、中段屋顶公园之间为庭院，使居民从公园中可
看到居委会中的活动。通过天桥跨过庭院可上到陈
列雕塑的公园最高处（雕塑仅显示基底轮廓）。

A courtyard between the east and middle sections
of the roof garden allows residents in the park to
look into the residents committee offices. Bridges
over the courtyard reaches the highest part of the
park, where two statues are located (only showing
the footprints).

公园中段的小广场可用作小团体活动。

The plaza in the middle section of the park can be
used for group activities.

公园游客可以通过公园中、西两段之间的庭院观看
社区活动室中的歌舞排练。

People in the roof garden can view the singing
and dancing in the community room through the
courtyard between the middle and west sections
of the park.

在高密度城市中创造公共空间
——昆山金谷园多功能建筑群（2013）

Creating Public Space in High-Density Cities
—Jinguyuan Mixed-Use Complex, Kunshan (2013)

问题

　　大多数十九世纪的中国城市没有多少有意规划的公共空间，特别是像广场或公园那样的节点空间，街道（或不如说是除去交通后的剩余面积）就是城市中的主要公共空间 [1]。1978 年以来，经济改革推动的城市改造彻底改变了城市面貌。但是，这一阶段的改造主要着眼于改善城市的经济基础设施，"无利可图"的公共空间没有得到相同比例的关注，特别是为普通居民服务的主体部分。所以，城市公共空间今天面临的第一个问题是数量上需要更多。同时，城市改造中已经提供的有限公共空间还存在三个质量问题 [2]。

　　首先，改善公共空间资源中大部分被用于少数几个"橱窗化"的政绩工程。这些巨型广场或绿地通常远离普通居民活动稠密的地方，由于中低收入的居民大多没有车，所以无法在日常使用这些设施，超大的草坪及烈日暴晒的硬地更是拒人于千里之外。其次，公共空间的"私有化"使商业街或社区中新建的少量公共空间也无法服务好普通居民。由于被经济效益至上的一些主管部门所忽视，这些"低效益"的工程通常能省就省，缺乏座椅、遮荫及其他必要的设施。与这些公共空间相邻的建筑立面经常是不透明的，拒绝内外交流。不少新建的公共空间甚至被商业化或被私人侵占。近 30 年来垄断居住区形式的封闭式小区，正在逐渐扼杀城市街道的公共生活。第三，公共空间日益趋向"贵族化"。新建公共空间中的免费或廉价的活动或服务大大减少，步行者及小贩经常被看作"二等公民"。

　　如果说在改革开放初期，以上这些问题可以被归咎于国家财政的窘困，现在已不再是这么回事。同时，越来越多的人正在往城市里搬，在 1978 年只有 17.9% 的中国人是城市居民，而在 2011 年上升到 51.3% [3]。这造成了越来越多的大城市。在 1985 年我国有 21 个百万以上人口的城市，到 2007 年这个数字已经飞跃到 117 个 [4]。与此同时，这些密集的城市人口又越来越中产化。所有这些趋势都要求我们必须尽快地改善城市日常生活环境，包括提供大量优质的公共空间。

　　但是还有一个更重大的理由要求我们创造更多更好的公共空间。今天，无论是公众、政府还是专业分析人士都已达成共识，那就是我国成功的经济改革要求对现有的社会和文化体系做与之相匹配的改革，像加强居民以主人翁的身份参与并管理社区公共生活，已经不能再继续推迟 [5]。要让更多的人参与到这场改革中来，我们必须加速推进群众自发的社区协商与文化交流。正如法国哲学家昂利·列斐伏尔（Henri Lefebvre）所指出过的，创造能容纳这些活动的公共空间是实现这一目的的重要条件 [6]。当然，我们必须警惕建筑决定论的陷阱，不能指望只靠物质环境就可以改变人的社会行为。但我们仍不能否认，是否有好的环境可以加快或拖延这一改变。现在是城市规划师与设计师为明天的公众参与创造摇篮的时候了。

　　如何在中国城市现有的约束中创造更多更好的公共空间呢？下面将用笔者最近的一个城市设计——昆山金谷园多功能建筑群为例，来解说一系列设计原则。

昆山金谷园多功能建筑群

　　坐落在上海西边的古城昆山已飞速演变成长江三角洲的一个主要工业城市。占 118 平方公里的市中心有人口 18 万，正在经历多项大型城市改造。本工程位于其中一个改造区域中。基地面积 1.38 公顷，呈条形，北边是刚完工的 5.6 公顷金谷园封闭式小区，南边是娄江，周围还有多片低层棚户住宅，将被改建为与金谷园（每公顷 155 户）似的高层、高密度小区。

　　考虑到金谷园小区北面就有一条主要商业街，建设单位要求将本工程建成一个建筑面积约 6 220 平方米的社区零售、餐饮及服务设施，包括一所含 6 个班级的幼儿园。公共空间要有，但未指定面积。其他人做的早期方案显示了时下习惯的手法，多为一长条一到二层的建筑，建筑与娄江之间剩下的窄条面积就算是"公共空间"。由于规范要求幼儿园必须坐落在自己的基地中，这些方案均将幼儿园放在基地西边的地面上。

八个设计概念

　　前面说的我国城市公共空间中现存的问题，部分是因为常规设计思路盲目照搬美城市形式，对中国城市的独特现状视而不见。这些状况包括：城市总体规模较大，现有公共空间有限（特别是节点空间稀少），夏天更热更长，大多数城市居民收入较低，以及需要保护城市周边的可耕地。但最不该忽视的一个国情是我国城市的高人口密度。举例来说，在 2012 年北京和上海中心城区里每平方公里各自有 7 387 及 16 828 个居民，远超过最拥挤的西方城市。高密度滋生了一系列其他问题，如对公共空间的高荷使用及更稠密的建筑密度。因此，我们必须有一套本质上与进口模式完全不同的设计对策。以下介绍的头五个概念就是主要针对高密度环境提出的 [7]。其余三个概念试图应对我国城市居民当前的经济及文化特征。

1. 大量小型的节点空间

　　在一个高密度城市中，众多袖珍型庭院或硬地比一二个超大"橱窗化"的公园广场要好。前者将形成一个公共空间网，使大多数居民在步行或骑自行车 500 米以内就可以到达一个公共空间。小型的尺度使得这类工程可以建在各种边角空地并减少拆迁。节点形态将创造较静止的环境，有利于形成社交聚会。传统中国城市不少庙宇或会馆中的庭院，实际上就是节点公共空间，它们弥补了街道作为公共空间的不足。在本工程中，按习惯手法连续铺开的一层建筑面积被缩小并分割成五栋建筑出地方来在各栋之间形成五个庭园。这些节点空间均有明确的边界，每个长宽在 42×34 米到 21 米 ×15 米。在公众眼中，它们将比仅仅加宽一点的街道更有效地鼓励们在这里开展社团活动。

垂直功能分区

要在稠密的城市里挖出更多的公共空间，一个办法是将不同功能叠加。传统的功分区通常为每个地块指定一种用途，比如是房屋还是室外空间，是这种建筑功能还那种。这种"水平"式分区不再能满足高密度城市的需要。在保证火灾及其他灾害会越过楼板的前提下，我们应当修改现行建筑及规划法规，允许垂直分区。本工程的某些功能已经如此做了。例如，A 栋的幼儿园中所有儿童用房连同室外游戏场地移到二层，只有主要为成人使用的服务供应用房留在一层的北边。整个幼儿园被设或一个小山村的形象，貌似山径的一个室外楼梯可被用作幼儿园的第二入口。这一布局使娄江景观不再被围墙阻挡，孩子们可以方便地眺望河景。它同时又改善了幼园的安全（当前群众关心的一个大问题）。但更重要的是，垂直分区使 A 栋一层的边得以成为商铺及一条公共敞廊，延续了从基地东面开始的娄江滨江景观带这一公步行道。建成后由于多种原因，建设单位最终将整个 A 栋（公共敞廊除外）交给幼园使用。但本实验证明垂直分区在设计及使用上是完全可行的。

3. 多层街道

由于人行道边的商店能吸引更多顾客，我国城市就出现了许多过度蔓延的商业街。这些街道不仅让购物者肉体疲劳，而且使城市中的区块难以被识别，特别是在大都市中。研究发现，长于 600 米的商业街难以吸引更多的顾客[8]。所以，我们为何不能通过在二层或半地下层设置人行道的方法来缩短一条街的商业部分呢？美国社会学家威廉姆·怀特（William Whyte）曾警告，只有靠近人流的公共空间才会被人使用[9]。因此，新的人行道离地面人流不宜超过一层，并应与地面通过频繁的公共楼梯连接。同时，地面上的行人应能看到二层人行道边商店的店面。多层街道的概念在本工程的 E 栋及B-D 栋中得到了局部应用。

4. 硬地花园

在高密度城市中，公园的本质是一个有绿化的公共大厅，而不再是一片移植到城市中的仅用于观赏的"自然"。时下公园设计的习惯做法喜欢采用英国景观式园林的模式，以大片草地、水面及树林为主。该模式无法支持公众对公园的高强度使用与损耗，这可以在许多国内城市公园中观察到。因此，公共绿地中的大部分地面应当是铺砌的，但同时设置尽可能多的位于上空或垂直面上的绿化。当然，可以设置少数几片面积有限、带保护措施的草坪，为使用者提供延伸视线的空间，同时也可满足偶尔的大型团

位置平面
　基地
　金谷园住宅区
　娄江
　白马泾路（规划）
　西仓基河
　震川西路

Location Map
Site
Jinguyuan residential development
Loujiang River
Baimajing Road (planned)
West Cangji River
West Zhenchuan Road

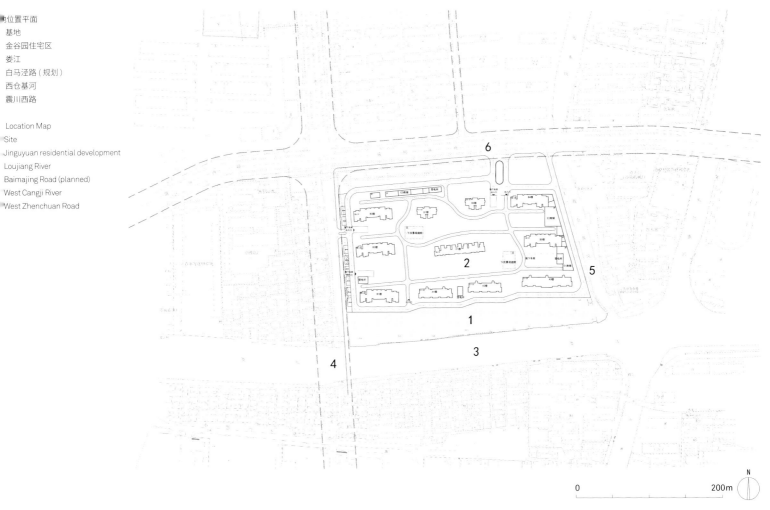

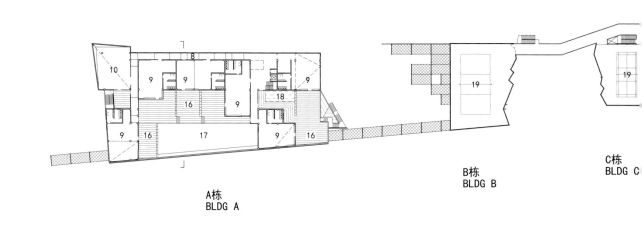

10

8

9 9

16

9 9

18

16 17 9 16

19

19

A栋
BLDG A

B栋
BLDG B

C栋
BLDG C

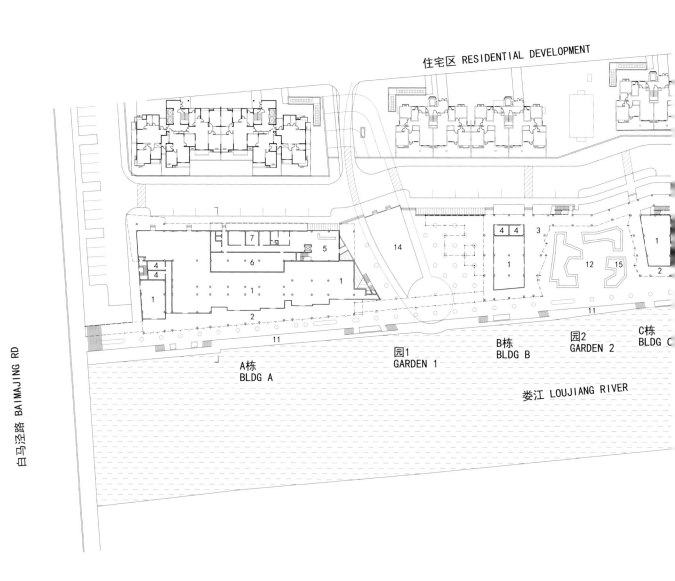

住宅区 RESIDENTIAL DEVELOPMENT

白马泾路 BAIMAJING RD

7

5

6

4
4

1

1

1

2

14

4 4

1

3

12 15

1

2

11

11

A栋
BLDG A

园1
GARDEN 1

B栋
BLDG B

园2
GARDEN 2

C栋
BLDG C

娄江 LOUJIANG RIVER

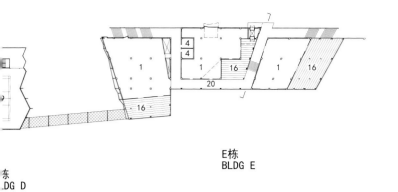

E栋
BLDG E

东
DG D

0   50m   N

（下）一层平面，
（上）二层平面
1   商店
2   敞廊
3   加宽敞廊
4   男／女厕
5   幼儿园门厅
6   幼儿园厨房
7   幼儿园办公
8   幼儿园坡道
9   幼儿园活动室 1-6
10   幼儿园音体活动室
11   台阶型河岸
12   小广场
13   台阶型平台
14   草坪
15   座凳环绕的种植坛
16   屋顶平台
17   幼儿园屋顶花园
18   幼儿园戏水池
19   屋顶球场
20   二层"街道"
21   步行桥

(Lower) First Floor Plan,
(upper) Second Floor Plan
1   Shop
2   Covered walkway
3   Expanded covered walkway
4   WC
5   Kindergarten lobby
6   Kindergarten kitchen
7   Kindergarten office
8   Kindergarten ramp
9   Kindergarten classroom 1-6
10   Kindergarten multi-purpose room
11   Multi-leveled riverbank
12   Mini-plaza
13   Stepped terrace
14   Lawn
15   Planter bordered with benches
16   Roof deck
17   Kindergarten roof garden
18   Kindergarten wading pool
19   Roof ball court
20   Second-floor "street"
21   Footbridge

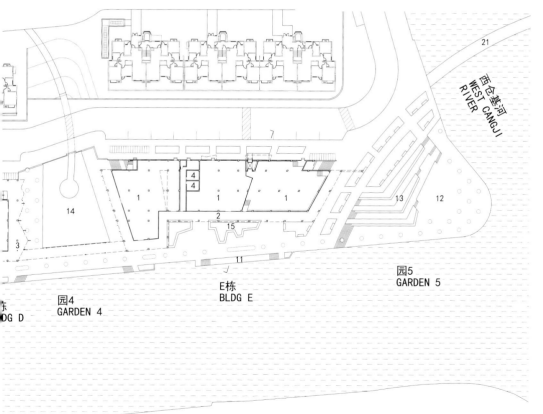

西仓基河
WEST CANGJI
RIVER

园5
GARDEN 5

园4
GARDEN 4

E栋
BLDG E

东
DG D

0   50m   N

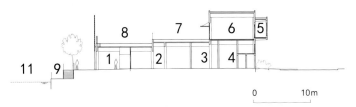

(左) A 栋剖面，   (Left) Building A Section,
(右) E 栋剖面   (right) Building E Section

1 敞廊	1 Covered walkway
2 商店	2 Shop
3 幼儿园厨房	3 Kindergarten kitchen
4 幼儿园办公	4 Kindergarten office
5 幼儿园坡道	5 Kindergarten ramp
6 幼儿园活动室 3	6 Kindergarten classroom 3
7 屋顶平台	7 Roof deck
8 幼儿园屋顶花园	8 Kindergarten roof garden
9 台阶型河岸	9 Multi-leveled riverbank
10 二层"街道"	10 Second-floor "street"
11 娄江	11 Loujiang River

体活动。公园应当能让人联想到大自然，但这必须用象征的手法而非模仿。以上这些想法均被应用于本工程中的五个庭园中。居民将在这些以硬地为主的花园里进行他们不大的公寓难以容纳的活动。大量的花架及树穴凸显了这些空间的自然本质。对应于北面金谷园小区的两个大门处，各设置了一片带围栏的小型草坪，开辟了通往娄江的视觉通道。在重要的节庆日，这些草坪可成为社区表演的舞台或孩子们的游戏场。

5. 硬质边界

　　在许多低密度的西方城市，领域感是用空间或甚至非物质的手段来产生的，前者像美国城市中独家住宅的宅前草坪，后者如西方文化的风俗习惯。但高密度的中国城市既没有过剩的空间也没有类似的文化（如很多国人排队习惯触及别人的身体），所以这些"柔性"边界行不通。目前常见的设计通常盲目模仿西方城市形式，所产生的无栅栏的草坪或其他空地很快就被破坏或占用。因此，我们必须采用物质的、特别是占用空间少的垂直屏障。在本工程中，矮墙、敞廊及高大的灌木丛在庭园与北面的街道之间形成一条"硬质"边界。在庭园中，高及人眼的密叶植物将各个交谈组团分隔开来，座凳同时是保护绿化的屏障。上面说到的两片草坪周边环有多处开启的活动栏杆，在偶尔需要使用草坪时可打开，使草坪与周边融为一体。

6. 低收费活动

　　以西方城市为楷模，我国城市规划者常为公共空间设想一些贵族化的使用方式，像交响音乐会或艺术家村。但事实上市民主体甚少光顾此类设施。一项 2012 年的调查发现，60.8%（最大群体）的上海市民使用最多的公共设施是简单的城市绿地 [10]。美国社会学家瑞·欧登伯格（Ray Oldenburg）指出，好的公共空间必须是免费或低收费的 [11]。我在上海的公园中也观察到，廉价的休闲方式最能吸引公众。除了人人喜爱的简易健身活动外，老年市民还喜欢棋牌及闲聊，中年人偏爱交谊舞及唱歌，年轻一代则沉醉于球类活动。为此，本工程中的五个庭园中设置了大量的座椅，其中许多被布置成向心型来促进安静的社交活动。庭园中同时设计了小片集中硬地来满足自我表现类的活动需要。考虑到社区公共空间规划常常忽视青少年的需求，B、C、D 栋的平屋顶被转化成一个排球场及两个羽毛球场。

7. 用途开放的公共空间

　　本工程中的五栋建筑均被一条至少 3 米宽的敞廊所连接，敞廊在有些段落的度还被进一步扩大。这些富余的尺度是为了让敞廊起到比交通走道更多的作用。为只有公众本身才知道所有使用公共空间的方式，最好的设计可能只需要提供一列物质条件简单但是免费的空间，让人们发明自己的使用方法，比如像英语角、鸟或露天咖啡座。增加一个屋顶可以用很少的成本激发最多样的使用。特别是，公共空间设摊是帮助低收入居民迈出谋生第一步的有效办法。管理得当的话，适的摊贩也能丰富公共空间。因此，本工程虽然规模有限，仍希望物业可以在敞廊中许少量的设摊。

8. 室内外空间配对

　　中国传统建筑的特点之一是将一个室内空间与一或多个室外空间配对来服务于一项主要建筑功能。与西方的大绿地中散布孤立建筑（如纽约的中央公园）的模式相比今天的中国人仍喜爱一种小块室内外空间混合的环境，既有室内的舒适，又与自然密切接触。本工程尝试了多种方式来继承这一传统，如交替布置建筑与庭园的总平面使庭园中的居民更方便地就近购买茶点，这也是欧登伯格所说的好的公共空间的一个特点。另外，A 栋中的幼儿园虽在二层，也为每个活动室提供了一个屋顶平台，有一个大型屋顶花园供所有六个班级玩耍。E 栋二层的每个商铺都有自己的屋顶平台与二层"街道"毗邻或可俯瞰娄江。

　　除了园 5 外，整个建筑群目前已基本建成。当本工程交付使用、攀援植物开始上来后，我们将可以看到以上设计概念是否可行。城市设计不能被简化为有趣的图设计，设计概念必须以人的行为模式为主要基础。但这甚至不仅是一个设计问题。如果附近的小区入口不按计划打开，就不会有许多人来使用公共空间；如果本建筑群商店中没有一个大众化的便利店，花园中的居民就不会进附近的建筑；如果物业不对小贩进行有效的管理，街市或是从不出现，或会惹得大家都讨厌。只有在一个社会中才会产生一个公民的公共空间，但我相信市民们一定能逐渐用好、管好自己公共空间——一个在中国漫长历史中刚出现的新事物。（2013）

释

] Pu Miao, ed., *Public Places in Asia Pacific Cities: Current Issues and Strategies* (Dordrecht, The Netherlands: Kluwer Academic Publishers, 2001), "Introduction," pp. 1-45.

] Pu Miao, "Brave New City: Three Problems in Chinese Urban Public Space since the 1980s," *Journal of Urban Design*, Vol. 16, No. 2, May 2011, pp. 179-207.

] 牛文元主编，《中国新型城市化报告》（北京：科学出版社，2012）。

] 刘乃全等，《中国城市体系规模结构演变：1985-2008》，《山东经济》2011 年第 2 期，第 5-14 页。

] 《2012 年"两会"温家宝总理答中外记者问 (实录) 全文》，中国网 www.china.com.cn，访问时间：2012 年 3 月 14 日。

] Henri Lefebvre, *The Production of Space* (Oxford: Blackwell, 1992).

] Pu Miao, ed., *Public Places in Asia Pacific Cities: Current Issues and Strategies* (Dordrecht, The Netherlands: Kluwer Academic Publishers, 2001), Chapter 13 "Design with High Density：A Chinese Perspective," pp. 273-293.

] 袁家方，《北京老商街的启示》，《商业经济研究》1996 年第 7 期，第 51-53 页。

] William Whyte, *The Social Life of Small Urban Spaces* (Washington，DC：Conservation Foundation，1980).

] 鲁哲，《公园绿地最受青睐，文化创意园光顾少》，《新民晚报》2012 年 7 月 26 日，A10 版。

1] Ray Oldenburg, *The Great Good Place* (New York: Paragon Books, 1989).

bstract

he existing Chinese high-density cities desperately need more and better public pace for the daily social and leisure activities of average residents, which are also ecessary for an emerging civil society. This design tries to experiment with the ght planning/design concepts proposed by the architect in the past, searching for pological solutions to the problem.

a renewal area of Kunshan's dense old town, the linear site sits between the ujiang River and the newly completed Jinguyuan, a high-rise, high-density (155 partments per hectare) residential development. The program calls for neighborhood ervice (including a kindergarten), retail and public spaces.

The design creates five small public gardens. Chinese cities lack nodal public paces which support social groupings better than expanded sidewalks. **Numerous mall Nodes** are better than a few large squares in a city because the former will be asier to be accessed by pedestrians and to be built on various lots.

Against the traditional planning of single-use lots, the **Vertical Zoning** of this design aces most part of the kindergarten on the top of a public covered walkway and ops so that urban land can be used more efficiently.

Over-extended on-grade commercial streets create fatigue and visual monotony. **Multi-layered Street** that is visually connected to the ground, such as the one in uilding E, can shorten the commercial street on the ground while utilizing a prime cation fully.

To support their intensive uses as public rooms, the green spaces in this project are ostly **Paved Gardens** with overhead or vertical planting. A small amount of lawn is ovided to allow for visual corridors and occasional group activities.

In a high-density Chinese city, one cannot rely on distance or etiquette to produce ffer between different functions. **Hard Edges**, such as wall and hedge that take less ace, should be used. The lawns mentioned above are surrounded by railings that n be opened during holidays.

Popular public spaces tend to support **Low-Cost Activities**. This design provides any centripetally-arranged benches, small "plazas," waterfront promenades, and of-top ball courts to facilitate conversations, social dancing, and exercises.

7. Several sections of the covered walkway in this project are expanded to provide **Spaces for Open-ended Uses**. Residents can invent their favored uses such as bird display or peddling.

8. **Buildings Paired with Open Spaces**, a tradition in Chinese architecture, is adapted in this design. The feature facilitates users of the public gardens to get food from nearby shops, a characteristic shared by many successful public spaces.

工程资料
地点 江苏省昆山市震川西路 (白马泾路与西仓基河之间)
时间 2009—2013
基地面积 1.38 公顷
建筑面积 6 220 平方米
业主 江苏省昆山城市建设投资发展有限公司
设计
建筑：缪朴设计工作室（缪朴）；汉嘉设计集团（蒋宁清）
结构：上海源规建筑结构设计事务所（张业巍、李明蔚）
设备：汉嘉设计集团（郭忠、于洋、吴秋燕）

发表
《Urban Design》（2013 年夏季刊），《建筑学报》(2013 年第 10 期），archdaily.com。

Project Data
Location West Zhenchuan Road (between Baimajing Road and West Cangji River), Kunshan, Jiangsu Province, China
Project Period 2009-2013
Site Area 1.38 hectare
Floor Area 6,220 square meters
Client Kunshan City Construction, Investment and Development Co., Ltd.
Designer
Architecture: Miao Design Studio (Design Architect), Pu Miao; Hanjia Design Group, Shanghai (Architect of Record), Jiang Ningqing
Structure: Shanghai Yuangui Structural Design Inc., Zhang Yewei, Li Mingwei
Engineering: Hanjia Design Group, Shanghai, Guo Zhong, Yu Yang, Wu Qiuyun

Publication
Urban Design (Summer/2013), *Architectural Journal* (10/2013), archdaily.com.

A 栋南立面
South elevation of Building A.

A 栋一层南边由一条敞廊连接的商店。
Shops linked by a covered walkway along the
southern edge of the first floor of Building A.

A 栋南立面及台阶型河岸。
South elevation of Building A and the multi-leveled
riverbank.

从 A 栋室外楼梯上到一个屋顶平台。
The exterior stair of Building A leads into a roof deck.

向西北方向看被幼儿园教室围绕的屋顶花园。
Looking northwest toward the roof garden surround
by classrooms in the kindergarten.

从活动室 1 平台下的矮空间内望屋顶花园。
Looking toward the roof garden from the crawl space
under the deck of classroom 1.

儿园活动室内景之一（室内装修由他人设计）。
e of the classroom interiors in the kindergarten
kerior decoration by others).

A 栋西北角二层为幼儿园音体活动室。
The multi-purpose room of the kindergarten on the
second floor at the northwest corner of Building A.

靠近 A 栋东北角的园 1 北入口。
North entrance of Garden 1, next to the northeast
corner of Building A.

园 1 中一片带可开栏杆的小型□
将小区大门引向娄江。

A small lawn, with openable fenc□
Garden 1 connects the commu□
gate to Loujiang River.

2，后为 B 栋。
rden 2 with Building B behind.

园 2，后为 B 栋。
Garden 2 with Building B behind.

B 栋东边缘面对园 2。
The east edge of Building B faces Garden 2.

B 栋东边局部加宽的敞廊，可满足公众的自发性用途。
The expanded area of the covered walkway on the east side of Building B, designed for "open-ended uses."

从 D 栋内望园 3。
From the inside of Building D looking into Garden 3.

（右）、D（左）栋的屋顶球场及两栋之间的园 3。
ll courts on the roofs of Buildings C (right) and
left), and Garden 3 between the two buildings.

D 栋前的园 4。
Garden 4 in front of Building D.

E 栋西立面，前为园 4。
West elevation of Building E with Garden 4 in its front.

被 D（左）、E（右）栋夹持的园 4。
Garden 4 flanked by Buildings D (left) and E (right).

D 栋东边缘面对园 4。
The east edge of Building D faces Garden 4.

E栋东北角为上到二层"街道"的室外楼梯之一。

Northeast corner of Building E, with one of the exterior stairs leading to the second-floor "street."

E 栋南立面。
South elevation of Building E.

E栋二层"街道"局部与屋顶平台相接。
The second-floor "street" in Building E is supplemented with roof decks.

透过 E 栋一个商店内的采光井可见另一面的二层"街道"。

The second-floor "street" is visible through the lighting well in one of the shops in Building E.

东二层"街道"
地面有视觉联系。

e second-floor
reet" in Building
has visual
nnections with
e ground.

E 栋二层"街道"与地面有视觉联系。
The second-floor "street" in Building E has visual
connections with the ground.

一扇诱惑人的大门
——昆山星溪公园游客中心（2017）

A Gate That Seduces
—Visitor Center, Xingxi Park, Kunshan (2017)

昆山市西邻阳澄湖，东与上海接界。从 1990 年代开始，它从一个有两千多年历史的古城快速转化为一个现代化工业制造中心。在这个过程中，城市也向周边不断扩大，即便是在远离中心城区西面 9 公里的阳澄湖畔，也可以看见成片的小别墅住宅区。记得那是 2010 年 7 月底，昆山城投的负责人把我找到湖边，向我描述了他要赶在私人住宅占满水边之前，在湖畔建设一系列公共休闲空间的计划。从昆山中心城区到阳澄湖有两条东西走向、大致平行的高速铁路通过（北为京沪高铁，南为沪宁城际铁路），他提出要在这两条高铁之间约 300~500 米宽的空地上建设一个长约 3 公里、面积约 140 公顷的线形公园。星溪公园在西端最终与阳澄湖畔的公园汇合，在东端则到高铁阳澄湖站为止。就在这个东端点上，城投邀请我设计公园的游客中心。

我觉得一个理想的游客中心首先应凸显该地的景观，让人对进一步探索产生兴趣，而不是用奇巧的外形来表现建筑自己。星溪公园的基本概念很有逻辑，就是把原来只能从经过的列车上扫一眼的所谓"绿地"废物利用，转化为城市居民周末的休闲健身场所。如果说前者是在地广人稀的美国大量高速公路旁常见的做法，后者则是对长江三角洲高密度城市寸土如金现状的创造性对策。星溪公园的景观设计也很朴素，不搞苏州园林或人工娱乐设施，而是简单地将基地上原来就星罗棋布的鱼塘、河道连接起来，形成一条以江南农村乡土植物为特色的湿地带。然后充分利用公园 3 公里长的线形平面，在其中布置多个成本低廉的慢行锻炼系统，包括步行道、约 7 公里长的自行车道、游艇航道，计划中还考虑了马车游览。这是非常切合中国现代工业化城市居民集体的需要的。

游客中心如何来向公众呈现这一片貌不惊人的乡土风景呢？基地现状为建筑设计提供了一个极好的条件。基地东面是沪宁城际铁路阳澄湖车站广场，两者被下沉的南北走向的新城路所分隔。其上建有一座步行桥，桥面离基地地面约 4.45 米。由于车站广场已经与附近主要道路的人行道、自行车道及多个公交系统有无障碍连接，我们因此将游客中心的主入口定在车站广场侧的步行桥头，而不是传统的基地周边地面上。这样游客进入中心时是在二层，正好为人们提供了一个居高临下俯瞰公园的好机会。设计还采用了三个先抑后扬的"悬念"处理来加强访客的感受。首先，73 米长的步行桥西端又增加了一条 18.7 米的延伸段，采用与步行桥相同的立面模数、地面材料及栏杆。人们在上桥前只能一瞥远处的游客中心，会产生某种好奇心，但他们必须走完这段 92 米长"走廊"后才能进入一个二层广场。广场周边用 3~5 米高的穿孔铝板及清水混凝土墙环绕，上空 6~7 米高处又覆盖花架，使广场成为一个半室外的"大厅"，游客暂时看不清楚周边环境。只在正对公园的西墙上开一扇近 9 米直径的半圆洞门，园内的景观到这里终于向游客展现自己。相信通过这两个前奏，即使是朴素的乡土景观，也会在游客眼中显得格外夺目。在洞门前是一个浅水池，它的反射水面好像铺了一条幻想中的入园道路。但实际上游客必须向右转，通过一个室外大楼梯下到一层，

那里才是多条入园道路的真正起点。这一短暂的等待，会在游客心中第三次产生悬念，增强他们对公园景观的向往。

建筑的一层大部架空，以方便公园保养车辆穿过建筑。游客中心的所有常规功能均在这里解决。一层西端为一玻璃圆柱体，内将用作导游展示及纪念品销售。在这里游客会发现刚才在二层看到的半圆洞门，其实是一个正圆形开口的上半部，它与导游空间的圆形平面形成了有趣的呼应。一层东端（步行桥延伸段的下面）将用作出租自行车的停车场，它有自己的借还车出入口。一层的中部则是木板墙面的公共厕所，被做成一个上部与二层楼板脱开、嵌套在一层空间中的独立体块，以强调二层的悬浮感。建筑西端悬挑在公园水面上，其南北两处水岸各设一游艇码头。建筑南面的地面广场则将用作未来的游览马车候车处。最后，建筑南北两边各与公园的步行 / 自行车环道相接，使游客中心成为线形公园中多条慢行锻炼系统的总出发点。

建筑两层的内容大不相同，是希望在满足功能需要的前提下，能为公众提供一点精神上的享受。近年来建成的不少游客中心把自己降低到商场、饭馆的层面，但我相信我国城市居民对公共空间很快将有更高的要求。本工程在试开放的几年中，看到有好多场婚礼自发地在此举行，而且其对空间的使用都采用同一顺序：新人从步行桥来，在二层广场的水池前拍照留念，然后从大楼梯下去。这说明只要我们把从设计到管理的工作做到家，公众完全知道如何创造公共空间的最好使用方式。

施工中有人问我，为什么要在二层广场的水池中设计一个内栽少许水生植物的花盆？我的意图是洞门及水池中的景观虽然好看，但尺度太大，与个体观众缺少联系。希望这几棵水草能让渺小的个人在这个宏大世界中也能找到他 / 她自己的位置。（2017）

Abstract

A three-kilometer-long linear park lies between two parallel elevated railways in the west suburb of Kunshan. For a high-density city, Xingxi Park converts a wasteland into a public space for urban residents' exercise activities such as hiking, biking, boating, etc. The site of the visitor center sits at the east end of the park. An existing footbridge over a sunken street connects the site with a railway station plaza, which is 4.45 meters higher than the site.

To arouse visitors' interest toward the wetland in the park, this design makes use of the existing site condition and the concept of "spatial suspense." Via the footbridge, visitors enter the visitor center at its second floor. Upon stepping on the bridge, people already have a glimpse of the distant park. However, they will not be in the park until they walk over the footbridge and its extension, enter a half-enclosed roof plaza, overlook the landscape through a semicircular gate behind a pool, and eventually descend into the park through a grand stair sideways.

With the second floor lifted on stilts to allow maintenance vehicles to pass, the first floor contains all utilitarian functions related to visitors' exercise activities. The contrast between the two floors suggests an intention to add a spiritual experience to the building type of visitor center that tends to focus on material needs. Since the opening, many wedding ceremonies were seen to take place in front of the gate and the pool.

Project Data
Location Xingxi Park, Xincheng Road, Kunshan, Jiangsu Province, China
Project Period 2011-2017
Floor Area 500 square meters
Client Kunshan City Construction, Investment and Development Co., Ltd.
Designer
Architecture: Miao Design Studio (Design Architect), Pu Miao; Shanghai Far East Architectural Design Institute (Architect of Record)
Structure: Shanghai Yuangui Structural Design Inc., Zhang Yewei, Miao Jianbo
Engineering: Shanghai Far East Architectural Design Institute, Song Yongchang, Wu Jun

Publication
New Architecture (2/2018), archdaily.cn.

工程资料
地点 江苏省昆山市新城路，星溪公园
时间 2011—2017
建筑面积 500 平方米
业主 江苏省昆山城市建设投资发展有限公司
设计
建筑：缪朴设计工作室（缪朴）；上海远东建筑设计院
结构：上海源规建筑结构设计事务所（张业巍、缪建波）
设备：上海远东建筑设计院（宋永昌、吴俊）

发表
《新建筑》（2018 年第 2 期），archdaily.cn。

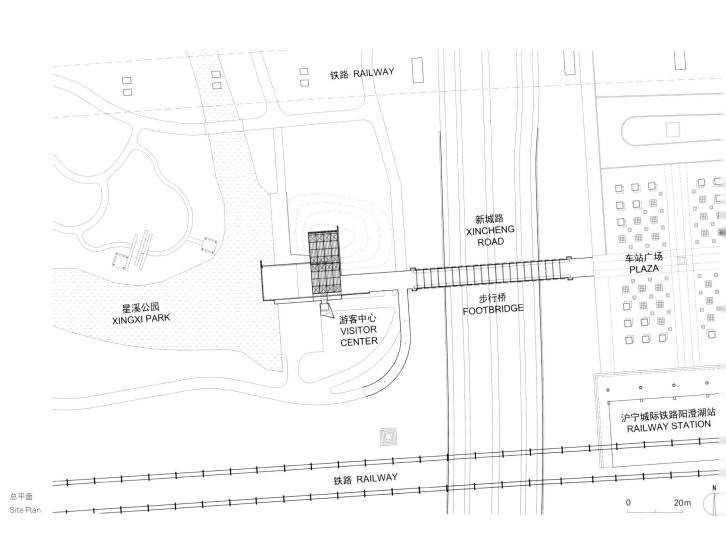

铁路 RAILWAY

新城路 XINCHENG ROAD

车站广场 PLAZA

步行桥 FOOTBRIDGE

星溪公园 XINGXI PARK

游客中心 VISITOR CENTER

沪宁城际铁路阳澄湖站 RAILWAY STATION

铁路 RAILWAY

总平面
Site Plan

0   20m

N

一层平面
1　商店
2　男 / 女厕
3　机房
4　管理
5　自行车出租
6　游艇码头
7　水面
8　广场
9　土坡
10　新城路

First Floor Plan
1　Shop
2　WC
3　Mechanical
4　Office
5　Bicycle rental
6　Boat pier
7　Lake
8　Plaza
9　Mound
10　Xincheng Road

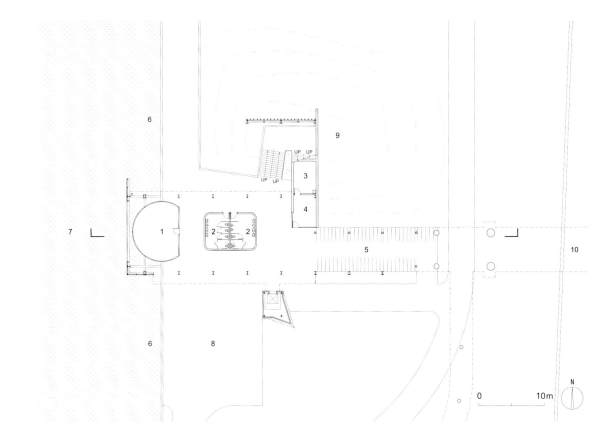

二层平面
1　现有步行桥
2　广场
3　浅水池
4　自行车坡道（后加）

Second Floor Plan
1　Existing footbridge
2　Plaza
3　Pool
4　Bicycle ramp (added later)

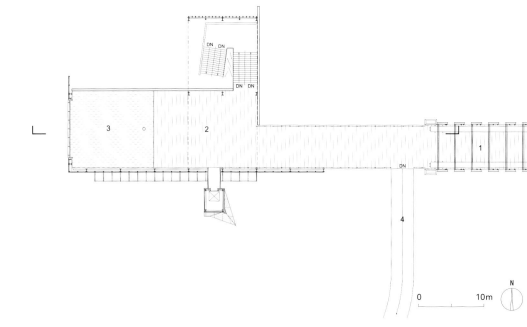

剖面
1　现有步行桥
2　广场
3　浅水池
4　商店
5　男 / 女厕
6　自行车出租
7　新城路

Section
1　Existing footbridge
2　Plaza
3　Pool
4　Shop
5　WC
6　Bicycle rental
7　Xincheng Road

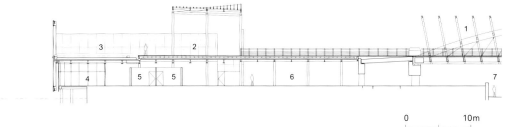

从车站广场沿步行桥西望，可一瞥游客中心的洞门。

Looking west from the railway station plaza through the footbridge, with a glimpse of the semicircular gate of the visitor center.

从现有步行桥西望。其延伸段采用了与桥相同的地面材料、栏杆等。远处二层广场半被混凝土墙遮掩。

From the existing footbridge looking west. The footbridge and its extension share the same paving and railing. The second-floor plaza is partially hidden behind the concrete wall.

从自行车坡道望建筑东南角。画面右边二层是现有步行桥的延伸段，连接绿色铝穿孔板墙后的二层广场。

Southeast view seen from the bicycle ramp. On the right is the extension of existing footbridge, connecting to the second-floor plaza behind the green perforated aluminum wall.

从现有步行桥延伸段西望二层广场。

rom the extension of the existing
ootbridge looking west toward the
econd-floor plaza.

层广场及浅水池，穿孔墙强调了
空间与精致感。

econd-floor plaza and the pool;
e perforated wall highlights the
nbiguity of grey space and a
licacy of the form.

二层广场西边的圆洞门展现公园中景观，浅水池好像铺就一条幻想中的入园道路。
The round opening of the second-floor plaza presents a commanding view of the park; the pool seems to lay an illusory path into the park.

从二层广场望大楼梯及花架。
From the plaza on the second floor looking toward
the grand stair and trellises.

二层广场上空的钢木花架，远处开口开向大楼梯。
A steel and wood trellis structure hovers above the
second-floor plaza; the distant opening leads to
the grand stair.

从大楼梯望建筑北立面西段。混凝土墙后是二层广场及浅水池。木墙面的公共厕所嵌套在架空的一层空间中。

From the grand stair looking toward the west part of the north elevation. Behind the concrete wall are the second-floor plaza and the pool. Nested inside the first floor are wood-sided public bathrooms.

建筑西北面外观，水边是游艇码头。

Northwest view of the building; along the waterfront are boat piers.

建筑西立面。
West view of the building.

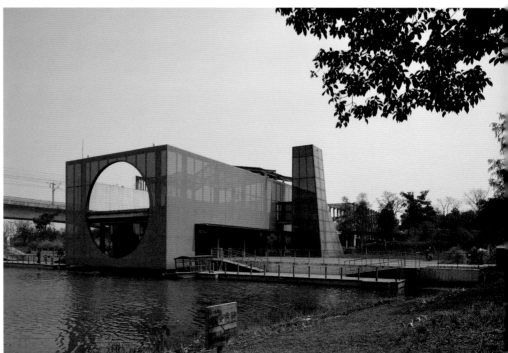

建筑西南面外观。
Southwest view of the building.

建筑南侧的电梯塔。
The elevator tower on the southern flank of the building.

建筑南立面西段。绿色铝穿孔板墙后是二层广场及浅水池。建筑前的地面广场将用作马车候车。木墙面的公共厕所嵌套在架空的一层空间中。

The west part of the south elevation. Behind the green perforated aluminum wall are the second-floor plaza and the pool. The open ground in front of the building will be used for boarding horse carriages. Public bathrooms with wood siding are nested in the first floor.

建筑西端南立面。左面玻璃圆柱体内将用作导游展示及纪念品销售；右为公共厕所。

The west end of the south elevation. To the left is the glass cylinder of the shop, to be used for tourist orientation and retail. To the right are public bathrooms.

架空的一层，远处为从二层广场下来的大楼梯。左为公共厕所。

The open first floor; in the distance is the grand stair leading to the plaza on the second floor. To the left are public bathrooms.

从游客中心一层东端西望。屋顶是现有步行桥的延伸段，连接二层广场。屋顶下的空间将作自行车出租。

From the east end of the first floor looking west. The roof is the extension of the footbridge, connecting to the second-floor plaza. The space under the roof will be used for bicycle rental.

圆形平面的导游空间与西立面上的
圆形洞门相呼应，将用作导游展示
及纪念品销售。

The cylindrical shop echoes the
round opening on the west facade;
the space will be used for tourist
orientation and retail.

二层广场及浅水池，穿孔墙使墙外
景观被部分渗透入墙内。

Second-floor plaza and the pool; the
perforated wall allows an elusive view
of the landscape outside.

其他设计作品的发表记录
Bibliography of Other Published Designs

上海浦东金桥新村中块一街坊规划（方案，1999）
No. 1 Neighborhood, Middle Block of Jinqiao Estate, Pudong, Shanghai (scheme, 1999)
缪朴，《封闭式小区：城市生活的"癌症"——问题与对策》，李磷、薛求理主编，《21世纪中国城市主义》（北京：中国建筑工业出版社，2017），第 166-187 页。

上海静安公园改建（方案，1998）
Renovation of Jingan Park, Shanghai (scheme, 1998)
缪朴，《"人造自然"还是"绿色大厅"——城市公园设计模式质疑》，《建筑师》第 91 期（1999 年 12 月），第 40-49 页。

深圳基督教教堂（方案，1998）
Shenzhen Christian Church, Shenzhen (scheme, 1998)
缪朴，《成对的室内外空间——深圳基督教堂方案》，《建筑学报》1999 年第 8 期，第 50-53 页。

上海三泉公园大门及餐厅（1997）
Entry Structures and Restaurant, Sanquan Park, Shanghai (1997)
缪朴，《用自己的声音说话——近作二则兼论"本土化"》，《建筑师》第 106 期（2003 年 12 月），第 20-28 页。

上海浦东新区中外宾接待中心别墅三栋（方案，1996）
Three Villas, Pudong International Reception Center, Shanghai (scheme, 1996)
刘尔明、羿风主编，《中国当代青年建筑师作品选》（北京：中国大百科全书出版社，1997），第 188-189 页。

园林设计两则（理论设计，1996）
Two Landscape Designs (theoretical design, 1996)
缪朴，《无限·另一个世界——园林小品两则》，《建筑师》第 73 期（1996 年 12 月），第 79-82 页。

深圳雕塑院（方案，1995）
Shenzhen Sculpture Institute, Shenzhen (scheme, 1995)
缪朴，《走第三条路》，《时代建筑》2000 年第 4 期，第 60-61 页。
刘尔明、羿风主编，《中国当代青年建筑师作品选》（北京：中国大百科全书出版社，1997），第 194-195 页。

深圳大学教工俱乐部（方案，1995）
Faculty Club, Shenzhen University (scheme, 1995)
缪朴，《成对的室内外空间——深圳大学教工活动中心方案》，《建筑学报》1996 年第 4 期，第 27-29 页。
刘尔明、羿风主编，《中国当代青年建筑师作品选》（北京：中国大百科全书出版社，1997），第 190-191 页。

深圳大学星瀚文化传播学院（方案，1994）
Xinghan College of Chinese Culture and Media, Shenzhen University, Shenzhen (scheme, 1994)
刘尔明、羿风主编，《中国当代青年建筑师作品选》（北京：中国大百科全书出版社，1997），第 192-193 页。

深圳儿童图书馆（方案，1993）
Shenzhen Children's Library, Shenzhen (scheme, 1993)
刘尔明、羿风主编，《中国当代青年建筑师作品选》（北京：中国大百科全书出版社，1997），第 196-199 页。

珠海伟大中心（方案，1992）
Weida Center, Zhuhai (scheme, 1992)
缪朴，《城市中心区建筑的四个规划概念——珠海某小区规划介绍》，《建筑学报》1994 年第 10 期，第 33-38 页。

作者简介
About the Author

缪朴（www.pumiao.net）是一位建筑师、建筑学者及教授，工作基地位于中国上海及美国夏威夷檀香山。他专攻现代建筑与城市设计在中国的本土化实践及理论。

注册建筑师缪朴从 1990 年代初开始设立缪朴设计工作室，探索创新的建筑、城市及园林设计。建成作品被国内外杂志发表，如《Architectural Review》《Domus》《Detail》《A+》《Urban Design (Quarterly)》《Architecture Asia》《Architectural Review Asia Pacific》《建筑学报》《时代建筑》《新建筑》及《建筑师》等。他的作品受邀参与了在柏林的 Aedes 建筑论坛（2011）及布鲁塞尔的国际城市、建筑及园林中心（CIVA，2008）举办的建筑展览以及第一届深圳城市 \ 建筑双年展（2005）。他获得的奖项中包括 2007 年的远东建筑佳作奖以及 2006 年的第一届上海市建筑学会建筑创作优秀奖。他的设计被现代中国建筑历史所讨论，如薛求理的《建造革命：1980 年以来的中国建筑》（香港大学出版社，2006）。

缪朴博士同时从事学术研究。他编著的《亚太城市的公共空间：当前的问题与对策》一书在 2001 年由克鲁尔学术出版社在荷兰出版，在国际上获得好评后又被翻译成中文。他是联合国人类住区规划署及意大利国家规划院召集的公共空间专家组会议成员（2014）。缪朴在国内外刊物上发表了大量学术论文，选题包括建筑与城市设计理论、园林理论、现象学、建筑教育、设计评论等。发表的刊物包括《Journal of Urban Design》《Nordisk Arkitekturforskning (Nordic Journal of Architectural Research)》《Landscape》《Places》《建筑师》《台湾大学建筑与城乡研究学报》等。他是最早对中国封闭式小区做专题研究（2003）的学者之一，其论文被许多国内外学者引用。他对中国 1980 年代以来的城市公共空间的分析，被选入城市设计理论读本。他研究中国传统建筑形式结构特点的论文被用作中国建筑学生的阅读材料。

缪朴在美国夏威夷大学建筑学院任教授，开设建筑设计及理论课程（课题包括建筑与城市设计理论、园林理论及电影与建筑）。他出生于上海（1950），在上海同济大学获得工学士学位（1982）后，又在美国加利福尼亚大学伯克利分校获得建筑学硕士（1985）及哲学博士学位（1992）。

Pu Miao (www.pumiao.net) is an architect, architectural scholar and educator based in Shanghai, China and Honolulu, USA. His work is dedicated to the design and theory of a modern architecture and urbanism localized in the context of China.

As a registered architect, Pu Miao established Miao Design Studio in the early 1990s, exploring innovations in architectural, urban, and landscape designs. His built designs have been published in international journals such as *Architectural Review*, *Domus*, *Detail*, *A+*, *Urban Design* (*Quarterly*), *Architecture Asia*, and *Architectural Review Asia Pacific*, as well as Chinese periodicals such as *Architectural Journal*, *Time+Architecture*, *New Architecture*, and *The Architect*. Miao was invited to exhibit his work at Aedes Architecture Forum in Berlin (2011), the International Center for Urbanism, Architecture and Landscape (CIVA) in Brussels (2008), and the First Shenzhen Biennial of Urbanism\Architecture, China (2005). The honors his designs received include the Design Merit Award in the 2007 Far Eastern Architectural Awards (Taiwan, China) and the Award of Excellence in the First Architectural Design Awards organized by the Architectural Society of China (ASC), Shanghai Chapter in 2006. His designs are reviewed by books on contemporary Chinese architecture, such as Charlie Q. L. Xue's *Building a Revolution: Chinese Architecture Since 1980* (Hong Kong: Hong Kong University Press, 2006).

Dr. Miao has also engaged in scholarly research. He edited and co-wrote the book *Public Places in Asia Pacific Cities: Current Issues and Strategies* (Dordrecht, The Netherlands: Kluwer Academic Publishers, 2001). The book received good reviews internationally and was translated into Chinese. He was a member of the Expert Group Meeting on public space convened by the United Nations Human Settlements Program (UN-Habitat) and the Italian National Planning Institute (INU) in 2014. Miao has also published numerous papers in international and Chinese journals on architectural and urban design theories, garden theory, phenomenology, architectural education, design criticism and other topics. The periodicals include *Journal of Urban Design*, *Nordisk Arkitekturforskning* (*Nordic Journal of Architectural Research*), *Landscape*, *Places*, *The Architect*, and *Journal of Building and Planning, Taiwan University*. He is one of the pioneers to study the gated communities in China (2003) and has been cited widely by international scholars. His analysis of Chinese urban public space since the 1980s was included in a reader of Chinese urban design theories. His research on the structural characteristics of Chinese traditional architectural form has been adopted as reading material for Chinese architectural students.

Dr. Miao is a Professor of Architecture at the School of Architecture, University of Hawaii at Manoa, USA, teaching architectural design studios and theoretical seminars (topics include architectural and urban design theories, garden theory and films and architecture). A native of Shanghai (born in 1950), he received his B.S. degree (1982) from Tongji University, Shanghai, a Master of Architecture (1985) and a Ph.D. degree (1992) from the University of California, Berkeley.

luminocity.cn

光 明 城

LUMINOCITY

"光明城"是同济大学出
版社城市、建筑、设计专
业出版品牌，致力以更新
的出版理念、更敏锐的视
角、更积极的态度，回应
今天中国城市、建筑与设
计领域的问题。

图书在版编目（CIP）数据

建筑与园林的对话：缪朴建筑设计作品集：汉、英 /
缪朴著 . -- 上海：同济大学出版社，2021.8
　ISBN 978-7-5608-9316-7

　Ⅰ . ①建… Ⅱ . ①缪… Ⅲ . ①建筑设计－作品集－中
国－现代 Ⅳ . ① TU206

　中国版本图书馆 CIP 数据核字 (2020) 第 107479 号

建筑与园林的对话
缪朴建筑设计作品集

缪朴　著

出 品 人：华春荣
责任编辑：晁　艳
特约编辑：杨碧琼　万月瑶
平面设计：张　微
责任校对：徐春莲
版　　次：2021 年 8 月第 1 版
印　　次：2021 年 8 月第 1 次印刷
印　　刷：上海雅昌艺术印刷有限公司
开　　本：889mm×1194mm　1/12
印　　张：18
字　　数：454 000
书　　号：ISBN 978-7-5608-9316-7
定　　价：198.00 元
出版发行：同济大学出版社
地　　址：上海市四平路 1239 号
邮政编码：200092
网　　址：http://www.tongjipress.com.cn
经　　销：全国各地新华书店

A Dialogue between Architecture and Landscape
Pu Miao's Architectural Design

by: Pu Miao

ISBN 978-7-5608-9316-7
Publisher: Hua Chunrong
Editors: Chao Yan, Yang Biqiong, Wan Yueyao
Graphic Designer: Zhang Wei
Proofreader: Xu Chunlian
Published in August 2021, by Tongji University Press,
1239 Siping Road, Shanghai, China, 200092.
www.tongjipress.com.cn